UNVARNISHED

ERIC ALPERIN AND DEBORAH STOLL

UNVARNISHED

A GIMLET-EYED LOOK
AT LIFE BEHIND THE BAR

HARPER WAVE

An Imprint of HarperCollinsPublishers

HarperCollins books may be purchased for educational, business, or sales promotional use. For information, please email the Special Markets Department at SPsales@harpercollins.com.

FIRST EDITION

Designed by Leah Carlson-Stanisic

Library of Congress Cataloging-in-Publication Data has been applied for.

ISBN 978-0-06-289928-6

20 21 22 23 24   LSC   10 9 8 7 6 5 4 3 2 1

FOR JONATHAN GOLD, WHOSE LOVE OF AVIATIONS,
SQUID SHIRTS, AND LATE NIGHTS CHANGED OUR LIVES FOREVER

IT IS SURELY A PROFESSIONAL RESPONSIBILITY FOR ALL OF US TO THINK DEEPLY ABOUT THESE THINGS SO THAT EACH OF US MAY BETTER STRIVE TOWARDS ATTAINING "DIGNITY" FOR OURSELVES.

—Kazuo Ishiguro, The Remains of the Day

CONTENTS

COCKTAILS AND DREAMS

They say I have talent.

Four years as a theater major at Mason Gross School of the Arts, in New Brunswick, New Jersey, and the world of professional acting awaits! But without a plan—or an agent, or any money at all—I find myself moving back into my parents' house in Westchester, New York, living in the very room I'd spent years excited to get the fuck out of.

My parents give me three months. After four, they fork over first, last, and security for a 187-square-foot apartment on Avenue D for $1,200 a month.

I promise to pay them back.

I register at a temp agency that places me in positions worthy of a college grad and spend my days sitting at ugly desks in ugly offices in ugly Midtown trying not to die of boredom. There is nothing for me to do in these jobs. I answer the phone once in a while. Copy some stuff. My job is just a placebo to make people feel like something's getting done when exactly nothing's getting done. I do my best to stop nodding out from monotony. Everyone thinks I'm a junkie, but I just need stimulation.

The same day the temp manager hands me a paycheck tucked inside a Betty Ford brochure, I spot an advertisement for a bartending school on the subway. I can't afford it, but my mother's sympathetic (more so than my father) and ponies up the deposit.

I promise to pay her back.

The two Italian brothers from New Jersey running the school teach me every drink in the lexicon of the National Bartenders School compendium: sticky, Technicolor eighties highballs and forgotten classics. They stress the dirty martini, the fuzzy navel, the grasshopper, Cape Cods, 7&7s, Jack & Cokes. I make notecards and study them with my mom. I pass. Place my National Bartenders School diploma on my wall next to my university BFA diploma.

I'm a double threat!

The brothers get me a catering job and I quit temping to spend my summer afternoons poured into overstarched white button-down shirts, highly flammable black Dickies pants, sometimes a clip-on bow tie, always a three-pocketed black apron. I stand behind makeshift tables draped in black tablecloths too low for any person over five feet to effectively work behind and make drinks for coked-up financiers throwing themselves parties. Pass hors d'oeuvres to eccentric art doyennes in SoHo galleries filled with "immersive art." Clean up after drunk wedding party guests floating up the Hudson. One night, bartending the after-party for *Henry IV* at Lincoln Center, I'm surprised and ashamed to run into a college classmate understudying the lead.

Fuck. Fuck. Fuck. I hate cater bartending.

I lie my way into a gig at a club called Cream, on the Upper West Side, where I start off barbacking on the floor. Move up to barbacking behind the bar. Finally I'm in my own well pouring drinks and realize I have every applicable neurosis for working in a bar: I'm really good at keeping credit cards organized. I can track who just walked in, who's been waiting forever, who cut the line, and who needs a fresh drink. I'm also super at keeping my tools organized, which helps me work fast, which impresses people, which makes them throw money at me, *which is awesome.*

I land an agent and get cast in a play off-off-off-off-etc.-Broadway. My GM, a former dancer, covers my shifts in exchange for working Sunday and Monday nights, which means less money, but this is my career we're talking about! I have to suffer for my art, so yes!

I take the shifts and the play is a hit and gets extended and my GM isn't so cool about it anymore.

I'm fired.

Making a total of $200 a week on an Equity contract for seven shows a week means I can no longer afford my apartment on Avenue D.

Shit. Fuck. Shit.

I call my parents, and they tell me to call my godparents, who tell me about a room to let on the second floor of their building on Washington Square Park. I move in with Milt Machlin—a septuagenarian who guzzles Gato Negro wine and eats chicken livers. The rent is cheap, but I still need a job, so I pad my résumé and get hired at the Screening Room, in Tribeca—a revolutionary movie theater/restaurant/bar concept. I feel totally at home behind the forty-foot bar slinging cosmopolitans,* which are on every cocktail menu south of Canal. The place isn't disco busy like Cream, but even when it's slow, it's interesting—I talk to the old man about his days as a Beat poet in San Francisco in the fifties; the single girl about her job as an alligator wrestler in Jupiter, Florida; the firemen from FDNY Ladder 8 about their closest calls. And

---

* Created in 1988 by then Odeon bartender Toby Cecchini, who was looking to create a version of a pink cocktail a waitress told him was all the rage in San Francisco's gay bars.

guess what? They respond when I ask them questions! It makes them feel important that I've singled them out and makes me feel purposeful, and before I know it everyone is talking to everyone else and I'm the star of the show.

One night, that classmate of mine from *Henry IV* comes in to meet a date and is pumped to see me slinging at this fancy joint. I buy them a round and feel like the leading man.

New York is a different city now that I know every bartender/barback/waiter/waitress south of Fourteenth Street—seafood towers at Blue Ribbon, cocaine and fortune-tellers at Raoul's, Sunday dance parties at Body & Soul—my new crew work at places I'd previously only read about and they usher me in, ply me with free drinks and drugs, and let me stay well past close.

I leave the Screening Room when I'm ready for something more serious and land at Lupa Osteria Romana, on Thompson Street. It's my first time working in "fine dining," and my mind is blown. I love the pre-service staff lineup, where we taste the dishes, recite their ingredients, and learn how to describe them to guests. I start mixing drinks with all the weird stuff behind the bar—grappa, amaro, kumquat cordials, and fresh citrus squeezed *à la minute*. But the wine flows heavier than the cocktails, and when I do get calls, they're usually for shaken-extra dirty Ketel martinis. Even when those calls come from Kathleen Turner's graveled alto, I know this isn't "it" for me. The problem is, I don't know what "it" is. Until the night I stumble into Sasha Petraske's Milk & Honey bar, on Eldridge Street. You know all those times you go out hoping to fall in love and don't fall in love because life doesn't work like that? I wasn't walking into M&H thinking of anything other than getting lit and getting loud, because it was supposed to be the bomb and it was my birthday. I didn't know M&H was a "classic cocktail bar." Shit, I didn't even know there was such a thing as a "classic cocktail bar."

But once I knew, I couldn't unknow. And when I fell, I fell hard.

Six months later, I wrangle myself a position at Little Branch—Sasha Petraske's new spot. I make enough my first month to pay my parents back. At home, I clock the diplomas on my bedroom wall. Realize my National Bartenders School diploma is worth way more than my university BFA. But that's just temporary, and right now I'm feeling pretty good. I'm working in one of the coolest spots in the city, my agent hasn't dropped me, and, to top it off, I have a date Friday night with a narcoleptic stuntwoman named Courtney.

CHAPTER ONE

NEW YORK, NEW YORK

It's Friday, 6:57 p.m. We have three minutes for last looks before the postwork revelers pour into Little Branch.

Located in the basement of a former transvestite bar on the origami triangle of Seventh Avenue South and Leroy Street, Little Branch is easy to miss. The exterior—a graffiti-covered white-brick wedge with an ugly brown door—looks like the illegitimate love child of the Flatiron Building and CBGB. The only way you'd suspect anything of note went on here was if you clocked the doorman standing unobtrusively outside, or if you got close enough to read the discreet silver plaque that says LITTLE BRANCH in the center of the door above the peephole.

Inside, a long set of vertiginous stairs lit only by flickering candlelight from below leads guests downward toward the clickety-clack of cocktails being shaken in tins. A standing bar cuts diagonally into the room and ends like the bow of a boat. To the right and left of the bow, five booths fan out along the wall, anchored in the center by a piano.

Starting at table 10, in the far street-side corner, I make sure each chocolate leather booth has been wiped with a damp bar rag followed by a dry one so when the first guests sit down, they don't sit on anything sticky or wet. Check that the teal wire-glass table-tops are fingerprint-free; fuckers show fingerprints like a crime-scene water glass. The oil lamps above the tables give off a glow that makes everyone look forever young, but the fixtures are old and precarious. We've jerry-rigged them time and time again. I do my best to secure the glass shades, but they'll be teetering on the edge of suicide in a matter of hours.

Whipping by the host stand, I count the laminated cocktail menus—a dozen—and restack them neatly on the first shelf.

At the bar, I glance across the stainless-steel garnish bins filled with crushed ice. Inside, whole raw eggs nestle on top, waiting to be cracked into flips and sours. Tiny silver bowls are filled with purple grapes and berries for smashing. Siciliana colossal olives and Luxardo cherries for floating. Lemon and lime wedges and orange slices for garnishing. Quartered limes, cucumber slices, and neat piles of mint leaves for muddling, plus sprigs for dressing. Running alongside the bin are eight eight-ounce stainless-steel pitchers filled with house syrups and fresh juices: simple, honey, grenadine, lemon, and lime. This is our mise en place, a French term common in restaurants that translates to "everything in its place." Since I'm half French and OCD, it makes perfect sense.

At the personality well,* I take stock that the jiggers are gathered together in a pyramid shape to the bartender's left and lined up in their respective measurement categories: six that have a two-ounce measure on one side and a one-ounce measure on the other; four that have a one-ounce measure on one side and a three-quarter-ounce on the other; and two that have a three-quarter-ounce measure on one side and a half-ounce on the other.

---

* The point of service behind the bar that requires the bartender to engage with guests who lean up for a drink.

My hand taps the four stirring spoons for martinis and manhattans and the two for cracking ice, submerged in a bain-marie* inside the crushed-ice bin. We keep all our perishables cold, but also our tools—cocktails should be served as cold as possible, and achieving this relies, in part, on chilled tools.

Is the muddler there? Check. It's resting gunslinger-style for quick draw by the bartender's right hand, just to the left of the spoons, and will be used to beat down citrus and lightly press mint in the stainless-steel eighteen-ounce cheater tins we build our cocktails in.

"Youse guys ready?!" Richard Boccato, our Italian-born, Brooklyn-raised host/door/sometime security man calls down the dark stairwell.

"S-s-stop acting the cunting maggot!" Micky McIlroy, our stuttering Irish bartender yells back. It means "Go fuck yourself" in Irish.

"Let's get flogged," Sam Ross, the inventor of the Penicillin cocktail barks in his Aussie accent. He switches out our prework jam—Social Distortion's "Ring of Fire"—for our early vibes playlist, and Eartha Kitt's "C'est Si Bon" fills the air as our first dozen guests descend the death-trap stairwell into our precious basement bar.

7:02 P.M.

"Party of four? Right this way."

"You're waiting on two more? Why don't you grab a spot at the bar and let me know when the rest of your party arrives."

---

* A cylindrical stainless-steel container traditionally submerged in boiling water to keep food items warm on a hot line. We use them to store our bar tools, submerging them in crushed ice to keep them cold.

"Greetings. Just the two of you?" I seat the couple at a deuce just to the right of the service station.

"Thank you," says the lady. "We're so excited to be here. We're visiting from Cleveland. We heard about this place from a friend. She said we had to come. I think she said we needed to order a *Park Sizzle?*"

"Yes, of course," I tell her. "You're referring to a Queens Park Swizzle. It's a version of a mojito. And for you, sir?"

"Surprise me with a Bartender's Choice."

Bartender's Choice was created at Milk & Honey where there was no menu. We have menus at Little Branch but incorporate Bartender's Choice as a fun option for adventurous drinkers. The way it works is, the server or bartender asks exploratory questions to figure out what a guest might like. It begins with "Do you have a spirit preference?"

"I like them all," my guest says, which either means he drinks a lot or is open to experimentation.

"Would you like something boozy or refreshing? Sweet or herbaceous? Aromatic or sour? Shaken or stirred? Long or on the rocks? Do you like eggs in your drink? Cream? Do you have any allergies we should be aware of?"

He stares.

"What about a style of cocktail?" I ask. "Is there a *type* of drink you usually like?"

"I love ginger!" he exclaims.

I slam down my first order of the night at the service well: a Queens Park Swizzle and an El Diablo.

"You sure you got that right Einstein?" Sam asks, picking up the chit.

"Uh . . . yes?"

"It's table *three*, not table two."

He shoves the ticket in my face and I see that in the bottom right corner where we place the table number I've put a "2."

"Shit, you're right man. The two guests made me think two."

I cross out the "2," insert a "3," then head back to the host stand to seat the next guests. The dual responsibilities of serving and hosting are a juggling act I'm still getting used to. I've been working here three months, but I've still got "new guy chops." Everyone else has worked in the family cocktail bars—Milk & Honey and East Side Company Bar*—and is able to tap in and out of any position any night of the week with ease, be it bartender, server, barback, or host. Unlike every other place I've worked, no one at Little Branch claims any single position. The idea is to be skilled at all of them, which allows everyone to think in a kind of "hive mind," aka group mind, aka gestalt intelligence, in which multiple minds are linked into a single, collective consciousness.

I seat the new party and head back to the service well, where Sam's got my first round ready, garnished and slowly dying. "At this rate you'll never make it to last call," he snickers.

"I'm good I'm good I got it," I assure him and whip the tray up into my hands, spin around without checking behind me and about-face into Daniel Boulud. Yes. Him. The famous French chef who, as much as he loves classic cocktails, probably doesn't love them on his shirt.

I wrest enough control of the tray to tilt it back, saving Daniel's blazer but completely hosing myself.

"I am *so sorry*," I tell him, Sam's carefully constructed cocktails dripping down my front.

---

* Sasha's cocktail bar on Essex Street which closed in 2010 and then became a wickedly dangerous but short-lived tiki bar called PKNY.

"Pas de problème," he cackles in his zippy accent, "but you, my friend, are soaked."

I look up to see Sam putting his hand over his heart—middle finger raised.

10 P.M.

Christy Pope, an M&H alum now working at Little Branch, shows up for triage. I thought I was handling things, but the room says otherwise. I'm behind on orders, have lost count of who's next and who's who on the wait list, and now I'm wafting *eau de gingembre.*[*]

As Christy helps me get the wheels back on track, I see Courtney, the girl I'm dating (we haven't had The Talk so I'm not calling her my girlfriend), posted up at the bar. We've been in Wynn Handman's[†] acting class together for a year, and falling for her was complicated by the fact that she had a boyfriend who was also in our class. I sometimes witnessed him being surly and verbally uncouth when we rolled over to P.J. Carney's Irish pub, on Fifty-seventh and Seventh, for all the Jamesons. One night when he wasn't in class, Courtney and I found ourselves squished up against a wall in the bar and I told her she deserved better, and she agreed, and we kissed. So I sought him out at his apartment to tell him how it was going to be.

"I already knew asshole," he said. "I can read behavior too you know." And instead of punching me, he punched the door and slammed it in my face.

---

[*] French for "ginger cologne."

[†] Legendary acting coach who was a member of the Neighborhood Playhouse and worked closely with Sandy Meisner from the Group Theater.

So now Courtney and I are dating.

When I try to catch her eye, I mis-garnish the old-fashioneds I'm supposed to be managing, which makes Micky grab the peeler to show me how it's done and he promptly slices off a piece of his index finger.

"Take over," he says under his breath, downplaying the blood pooling fast in his palm.

"What?" I ask, my stomach flipping, because, while I feel for-midable on the floor and capable behind the bar, I'm definitely not ready to be back there weekends.

"And remember Romeo," Micky says before disappearing, "when you jigger, make clean and defined movements, and after you pour the spirit, say a little prayer before dumping it into the tin."

Sayruf, our eternally good-natured Bangladeshi barback who notices *everything*, slides in behind and loads me up with clean tins, then checks to make sure the garnish bins are full and that I have all the ice I need.

I got this, I tell myself as Christy, who looks at me questioningly but doesn't take the time to ask questions, throws down one ticket for a Blue Collar, which is a manhattan twist, and another ticket for four drinks: a Gold Rush and an Enzoni, which are shaken and served down, a Moscow Mule, which is shaken and served long, and an Eastside, which is shaken and served up.

"Christy?" I point to the Blue Collar.

"Orange bits, quarter Maraschino, quarter CioCiaro, half sweet, two rye, lemon twist," she says. "It's Madrusan's cocktail—remember? We made it together on our Wednesday shift."

"Oh yeah, right. Shit. I got this."

I got this I got this I got this.

I set up a chilled mixing vessel and four tins from left to right, which is the order the drinks are written on the chit and also the order in which the guests are seated, then start with the cheapest ingredients first so if I mess up early on in the process, I'm not dumping the gold.

Muddled items first (mint, cucumber, grapes). Then bitters, syrups, and citrus. Then the amari,* liqueurs, and booze.

I remember to jigger left to right, doing my best to reuse the same jigger for multiple ingredients until no others can be crossed. Lemon and lime can use the same jigger, and simple syrup can go before honey, or any syrup for that matter, but not the other way around. Vodka can go before any other booze, and light rum can be followed by dark rum, like rye before bourbon. The whole idea is efficiency and execution so that the first finished and last trayed cocktails are completed within thirty seconds of each other, delivered and served to a table at the same time. It's like steaks ordered at different temperatures all landing on the table concurrently—everyone should eat together; everyone should drink together.

I tray the first round, which Christy whips up into her hands and takes off into the room with, and start on my second. I can feel myself gaining confidence, sliding into the zone, my movements syncing to the music—Al Foster's pre-noise-rock cymbal crashes on Miles Davis's "Moja." It was recorded live at Carnegie Hall, so it sounds like Miles and his band are here in this room. I glance up. The whole room's watching. Well okay not the whole room, but it feels like it, except instead of making me nervous it steadies me.

I twist the cap off a bottle and flip it under my middle finger. It stays in place as I grab a jigger, pour, and spin the cap back on. Quarter turn to place the bottle back on the shelf, label side out, and pull off another just like that. I'm not thinking; only a

---

* Amari (singular: amaro), which means "bitter," is an Italian herbal liqueur typically consumed after eating to help with digestion. You know Jägermeister? That's amaro from Germany. Same same; different country.

mnemonic refrain for the drinks runs through my brain: *MM* for Mercy, Mercy; *SR* for Southside Rickey; *CC2* for Clover Club No. 2; *Enzo* for Enzoni.

MM, SR, CC2, Enzo. MM, SR, CC2, Enzo.

I pop the four drinks up onto a tray and am working through my third round, a full jigger in hand, when Sam sneaks up behind, reaches around, and dick-taps* me, causing me to spill gin across the tops of all the tins.

"Oh, sorry bud. You gotta start that round over," he laughs. "I'm bouncing to M&H to help cut ice. Cervantes will be here in five. Christy's got the floor handled, unlike you, but no worries pal. You'll get there."

11:17 P.M.

Sweat drips down the center of my back as the air from the old HVAC vent stationed directly across from the service well works overtime trying to keep the bar area cool, but it's stacked three deep, guests' winter layers draped across their arms or hanging on hooks when they can find them. Bodies backed up to the stairs. Chits are piling up and guests are staring at me. "Hold On, I'm Comin'," by Sam & Dave, plays as Cervantes enters the scene. The chillest Dominican man I have ever had the pleasure of bartending beside. He parts the crowd like Moses and the Red Sea and floats behind the bar, tagging me out.

With the bar and floor covered, I race upstairs to the staff hangout, which is supposed to become a private event bar at some

---

* A light tap of a man's private area, usually done with the knuckles.

point,* to devour the Grey Dog sandwich I bought five hours ago which is now soggy. I'm wolfing it over a trash can, juices dripping down my arm, when Courtney appears at the top of the stairs.

"Nice table manners Sonny Bono," she says.

I assume she's insulting my peach polyester button-up and acetate baby blue blazer, finished off with a sailboat tie courtesy of the Salvation Army on Eighth. She either really likes me or takes pity on me or some combination thereof, because in all the tumult, I totally forgot she was in the bar and have been ignoring her for hours.

"Thanks," I smile. Hoping there's nothing in my teeth. "I'm sorry I couldn't talk . . . It got . . ." I gesture toward the bar downstairs.

"Don't worry about it," Courtney says. "You looked pretty good back there. Seems like the hell Sam is giving you is only because he cares."

"Yeah," I say. "Tough love. I hope you're right."

"Well . . ." She edges toward the exit. "I just wanted to say goodnight. I've got rehearsal early with that L.A. director."

"Oh yeah," I say. "That's so cool. Good luck. Which you don't need because you're great."

"Thanks," she smiles, which makes her eyes crinkle, which makes me sweat. "I'll call you after."

When she leaves, I wonder if she likes me as much as I like her. If I'll ever have a rehearsal with an L.A. director or if I'll be eternally shoving soggy sandwiches down my throat in bars at midnight.

---

* It never does.

1:06 A.M.

The lovely couple from Ohio stay for a few too many and need help getting up the stairs. "Here's our number if you ever open a bar in Cleveland," reads the note they leave on the table.

Two brothers from Paris at table 2 fall in love with the ladies at table 3 and ask me to introduce them. Within the hour I've moved them all to a four-top.

I communicate with a Japanese couple visiting from Tokyo via their map of Manhattan, circling neighborhoods and attractions I think they might like, bring a round of amaro to a table of locals, and tell the beautiful girls at table 6 playing Truth or Dare that while I can't meet them later at Kokie's, maybe ask Cervantes? . . .

3:45 A.M.

"How'd it go tonight?" Richie asks, raising the "fuck-off lights" to full blast.

"I'm still here," I tell him.

"Gonna f-f-fire you tomorrow," Micky yells from upstairs, his mouth full of Grey Dog. "This sandwich sucks!"

"Hey, you think I should give working the bar a shot?" Richie leans up against the bar and takes a sip of his shift drink—Pappy Van Winkle Family Reserve 13-Year Rye. His position at the door is the only one singular in nature, since it's at the top of the stairs and not a role you can tag in and out of during service.

"Fuckin-A yes you'll kill it brother," I tell him. "Just work left to right, cheapest to more expensive. Oh, and remember that a Pink Lady doesn't get ginger syrup. I screwed that one up twice tonight."

"Yeah. Cool. Pink Ladies. Thanks man."

"Come on Richie!" Micky yells down. "Let's go."

"Meet us at Blue Ribbon when you're done," Richie says, gulps down the rest of his drink and takes off.

The rest of close-out duties fall to me, Cervantes, and Sayruf, and we start ripping the bar down piece by piece. All the custom stainless grates are Tetrised together, and even if two pieces look identical, one fits in a spot of the scupper drain better than the other. Most of the grates are too large to fit in the glass washer, so we hose and wipe them down by hand. This part of the night's work is what I think people mean when they say something's "meditative." My body, which has been held tight at attention all night relaxes, my shoulders fall away from my ears, my feet move with purpose but less hurried now. The whole vibe of the room is like this, all of us in a dance but slower—not a *slow dance*, but no longer moving with such intensity. A lot of people hate closing and cleaning, but to me it's a perfect coda, a time to mentally take stock of the night as I clean and polish and replace and replenish.

Attacking the garnish bin, I dump the night's fruits and juices, because at this point they've lived through the open air of a room filled with bodies. It's not much waste, since Sayruf takes the orange juice home, and the grapefruit I defer to Cervantes because I'm still the new guy and looking to shore up my relationships.

Each garnish bin has a rectangular piece of wired glass trimmed in stainless, which slides into its front and acts as a kind of sneeze guard. More important, it deters guests from reaching in and grabbing pieces of fruit. After a night of service, the glass is cold and slippery. I do my best not to crack it, using kid gloves to wipe it down before placing it gently on the bar top to dry.

I lift up the over-sink stainless-steel adjustable strainers inside each dedicated dump sink, flip them over the trash can to knock out the muck of dead muddled ingredients and tasting straws,[*] then use a paper towel to pick them clean before passing them off to Sayruf for the wash. Jiggers, spoons, muddlers, and dirty tins get hauled over to the pile of items awaiting their own spin.

The speed-rail bottles get buffed with a rag, then I do the same with the backbar bottles, which I also twist so they face label side out. The practice of cleaning and wiping and facing as we go is ingrained in us all, so none of the work I'm doing is major. It's mostly just tweaking and making sure everything is ready for tomorrow night; that the next bartender is set up out of the gate.

Back out on the floor, I gather any menus that have been left out, wipe them down and place them in the host stand, extinguish any oil lamps and votives still lit, give them a buff so the paraffin buildup doesn't grease their sides, then throw all the drinks left on tables onto a tray.

"Stop. Go drink beer," says Sayruf with a toothy grin. It's a kind way of telling me I'm overloading the dish line.

"Sorry," I mumble and reach into the cooler, grab two Czech Rebel lagers—one for Cervantes and one for myself (Sayruf doesn't drink)—and sit down in a booth to start counting money. Cervantes fires up a joint, taking huge puffs that hang heavy in the still barroom air. Then, like a drug-counting assembly line, we place bills in their respective piles facing up and looking left, count out tomorrow's bank—$400—and fill in the close-out sheet which asks for gross sales, individual staff hours, and tips as well as the cash drop, then rubber-band all the money and drop it into a Carpano Antica vermouth tin that's been repurposed as a bank bag. I pop it into the far left corner of the bar by the espresso machine on the

---

[*] Straws the bartender uses to taste cocktails to ensure they're balanced.

lowest shelf behind some backup booze—our version of dropping money into a safe.

"Your turn guys," Sayruf tells us, having left the floated wooden floorboards, aka "deck boxes," for us to haul out from behind the bar. Standing on a deck all night, as opposed to hard concrete, can seriously extend a bartender's life. They also sit a bit higher off the floor, giving us a better sight line of the room. The downside is they're heavy as fuck.

Cervantes and I tip each deck together, one by one, and lean them along the customer-side bar wall, leaving the floor clear for the porters to come in and clean. The final job of the night.

We stand still for a second under the bright work lights. It's something beautiful now, like standing onstage after a show with the audience gone and you can see the spike marks* on the floor and the glue holding everything together. Seeing places in ways they aren't meant to be seen makes me feel like I've been given the keys to something special.

AT 5 A.M. WE LOCK the front door. Some nights, all you want to do after work is crawl into bed. Others, you're too amped for sleep and want to dance, drink, or socialize until dawn.

Sayruf decides to head home and hops into a cab. Cervantes takes off on foot to some chick's place, probably someone he charmed earlier from behind the bar or the girls who were headed to Williamsburg. I stand alone in the cold, weighing whether I'm up for meeting Micky and Richie and the rest of the industry owls at Blue Ribbon. Across the street, Daddy-O's stragglers are getting kicked to the curb and I'm tempted to pop over to enjoy a lock-in with the

---

* A mark, usually made with a piece of tape, that shows an actor where to stand.

crew. But my bed calls. So I pull up my hood and head east on Carmine, cross Sixth to Minetta Lane toward Macdougal, where a line of drunks are spilling out of Mamoun's—the best $2 falafel ever served out of a closet-size kitchen. The smell is a siren call, and I hop on the end of the line. Thankfully, I don't have to wait long—the more inebriated patrons wander off—and soon enough I have a warm fried sandwich in hand and am wending my way home.

"Hey man, you need green? Got green."

I'm always propositioned for weed in Washington Square Park. Sophomore year of high school, I was forever taking Metro-North from Westchester to buy marijuana for me and my friends and learned some big lessons about seeing the goods before giving the cash. I got ripped off twice, nearly swearing off drugs before a couple of NYU stoners took pity on me and gave me a free dime bag of shake. And a dime bag of shake, once you pick out all the seeds and stems, goes a long way among high school kids.

"Whatcha got man?" I ask.

"Good green. You'll like it."

"Gotta see it first."

He pulls out a bag.

I shake my head and frown. "I don't like oregano," I tell him.

"Ain't no fuckin' oregano, bumbaclot man!" he yells into the wind. It would have been nice to get stoned, but I don't need it to sleep. Between the walk, the food, and the cold, I'm more than ready to peace out. I land on my stoop at 27 Washington Square North, reach into my pocket for my keys and . . . *Where the fuck are my keys?*

I pat myself down, thinking I just can't feel them through the bulk of my layers, and then recall taking them out of my pocket and placing them behind the cash drawer when I got behind the bar so they wouldn't bite into my leg as I leaned against the well. My hand hovers over my buzzer, but my seventy-three-year-old roommate, Milt, who wrote a book about Michael Rockefeller mysteriously disappearing in New Guinea and allegedly getting eaten by cannibals, is unlikely to wake to the sound. Besides, I don't want to startle the guy.

I'm starting to feel supremely cold as it dawns on me that I can't even get back into Little Branch to get my keys because my *bar* key is also on my *key chain* which is *inside* Little Branch and I'm about to sit down on the stoop and sleep right there when I remember that Sasha lives in the staff room.

I hail a cab, and five minutes later am knocking gently on Little Branch's steel side door.

Nothing.

Knock a little harder.

Nothing.

I turn back to face the street with the depressing realization I'm going to have to take up real estate at Veselka until Milt wakes up when—

"Eric?"

I turn around to see Sasha cracking open the side door.

"Oh Sasha. I hope you weren't sleeping. I forgot my keys."

"I was just making myself an old-fashioned," he says. "Would you like one?"

I follow him gratefully inside and downstairs, amazed at how for a broad man, he moves with such grace.

All the lights in the room are off except for the spots over the bar. Sasha walks behind the service well. He's undone his tie and top buttons, loosened the grips of his suspenders.

"I'm only making old-fashioneds as we don't want to disrupt the fine work that Sayruf does for us," he explains as he sets out his ingredients. "You guys are lucky back here. I hope he never leaves us."

"He's the best. Especially to me. I'm still, you know, struggling with rounds and he always has my back."

"You'll get there," Sasha assures me. "Takes time, but then one shift . . . it all clicks."

He takes out two rocks glasses, drops a sugar cube in each, adds three dashes of bitters and a touch of soda. He muddles each into a paste, pours in the Elijah Craig bourbon, then grabs two rocks of ice and gently lays one in each vessel.

"Mind peeling some citrus?"

"Of course," I say, joining him behind the bar.

"Be careful when you peel the oranges to leave the least amount of rind on the underside," he cautions. "It should look like the skin of a Band-Aid, where you can see all the dimples."

"Okay. You got it. Appreciate that."

He finishes off the drinks and slides one my way.

"Since you're an actor, you'll probably book something big and leave us for Hollywood, won't you?" he asks, gazing into the middle distance.

"I don't think so," I tell him. "I like theater, which is in New York, you know?"

"I have a cousin in Los Angeles," he says, like he didn't really hear me. "It's always sunny there and people seem healthier. And with all the colors you've chosen for your outfit, all that's missing is a palm."

"I'm just having a little fun."

"Of course, but may I suggest a different material? Something that breathes, perhaps?" Sasha takes a long sip of his drink. Then he picks at the orange peel, inspecting it. "Can you imagine doing yoga in L.A.?"

"I've never done yoga," I tell him. "But I do like to drive, which I know is strange for a New Yorker to say."

"I don't even have my license," Sasha says, then pauses. "What kind of car do you like?"

"A convertible, for sure."

"Yes," Sasha smiles and raises his glass to mine. "A convertible. I'll buy an extra case of Royal Crown* so I don't muss my hair."

---

* Hairdressing pomade.

CHAPTER TWO

ROOT DOWN

My third week in L.A. and I'm zipping along the PCH* in a beat-up red Nissan 300ZX with the T-top panels removed. Ocean breezes, girls and boys on roller skates, girls and boys on skateboards, girls and boys wearing bathing suits in public like the city is one giant beach—it's all happening! I'm on hold for a contract role on *The Young and the Restless*—not Scorsese, but it's a start! And as my agent reminds me, some of our finest actors starred in soaps, so I'm in good company.

As it happened, Sasha had been contemplating opening a bar in Los Angeles for a while and was testing the waters with me the night I got locked out. A week after our late-night session, he asked if I was interested in partnering with him, which made me feel the same as when Allison Greenhouse agreed to be my date for prom. Elated! Incredulous! Terrified!

When I bounced the idea off of Courtney and she said, "I think it's a good idea; I'll move with you," I packed my bags, bought a car, and drove cross-country to begin life anew.

---

* Pacific Coast Highway.

◦→▬◉ ◉▬←◦

MY PHONE IS RINGING. IT'S her! My agent!

"Honey darling Eric how are you?"

She speaks without punctuation—she has no time to pause.

"Great!" I exclaim, wondering what's smarter—investing in a house in Malibu or making my commute to the television studio easier and buying a place in the Hollywood Hills.

"About the test deal—" she says.

"Yes!" I exclaim, wondering how to break the news to Sasha that I can't open a West Coast branch of Milk & Honey because I have to be a soap star.

"You were fabulous—" my agent coos. "Wonderful they loved you—"

The sun glints in my eyes as I relive the moment of improv no one in the audition room was expecting. I *knew* it was right, that the script was cool, but what they were really looking for was *personality*. Someone willing to make shit up! The anguish I feel over disappointing Sasha is alleviated by thoughts of the money I'll have from the show. So much money I can invest in his bar with whoever he opens it with. And visit! In bespoke suits!

"Honey darling Eric they said it came down to just the two of you but in the end they decided to go with—"

A seagull screeches.

I swerve to avoid a homeless guy jogging across the PCH carrying a surfboard.

A wave crashes off the rocks to my left and the name of the guy who was better than me or shorter than me or beefier or funnier or quirkier or whatever-ier he was that landed him the job instead of me is lost forever to the pounding surf.

I'd believed in the adage "If you can make it here, you can make it anywhere," and assumed L.A. would be a breeze. But I hadn't "made it" in New York at all, and my off-Broadway résumé, sprinkled with a few guest-starring roles, wasn't impressing anyone. The longer I'd gone without getting cast, the more unaccomplished, nervous, and unsure of myself I'd become, which is kryptonite to an actor who needs all the confidence in the world when walking into a casting director's office. Add to that the fact I haven't found any bartending work and I'm feeling like a total loser.

I pull into the narrow parking lot at the top of Sunset Beach and wedge my car into a space behind a Vanagon. Get out and look over the sea-stack rocks, slick with salt water from the waves crashing against the shore below. A woman sits on a towel on a thin sliver of sand, staring up at the homeless man shimmying into a wetsuit, and I realize he isn't homeless at all. He's a surfer. She blows him a kiss as he jogs to the shoreline, and at the water's edge, in a single fluid motion, glides effortlessly onto his board and begins paddling toward the lineup. Things are never what they seem: the homeless man is a fit surfer dude with a hot girlfriend. I think I'm an actor, but reality says otherwise. How many more phone calls that go, "Thank you, that was great, we appreciate you coming in but we've gone in another direction" can I take? It hasn't been that long, I know, a blip on the radar, but the more time I spend talking about opening a bar, about creating my own job and jobs for other people too, the less down I am with waiting around for someone to tell me when I can and can't "perform."

I really wanted to have my shit set up when Courtney arrived. She stayed in New York to shoot a movie and is set to fly out here next week—what the fuck is she going to think when she moves to California to find her boyfriend unemployed, suffering night sweats in the shithole apartment I found for us on Fairfax and Third, and seriously considering going back home? I'm thinking maybe I should call and tell her this was all a mistake when my phone rings and I pull it out of my pocket without looking at the number and dramatically exhale—

"Hello?"

"Hi, is this Eric?" asks an unfamiliar voice.

"Yeah."

"Hi, this is David Rosoff from Osteria Mozza. We'd love to offer you the bar manager position."

It isn't TV, but it's a job.

MY FIRST DAY AT MOZZA, a box truck is backed up to the employee entrance. A stylish, petite woman in a colorfully patterned jumpsuit and round red sunglasses stands on the lift gate, holding a box of leafy greens.

"Hey. You. Who are you? Can you help bring these to the kitchen?"

This is how I meet Nancy Silverton, who speaks in staccato phrases, crafting her sentences as if weighing each word before releasing it into the world. For the next fifteen months, my knowledge base expands beyond booze and brands and cocktail specs, to the actual business side of things. Thanks to the GM, David Rosoff, I learn how to prep inventory, create par sheets,[*] and manage costs, margins, and my team. Nancy's business partners—Mario Batali and Joe Bastianich—solicit my help curating the backbar amari, which are sent over from Italy by a long-haired hippie who drives from the top of the boot to the bottom and out to the islands in a VW bus collecting the liqueur. These beauties adorn our backbar— more than a hundred different bottles—which we pull down when we want to pour something special.[†]

---

* A reckoning of the amount of goods you have on hand to sell.

† They are never sold.

But my most significant lessons at Mozza have nothing to do with the bar, or at least not the one with liquor behind it—Nancy's mozzarella bar is a shrine to soft, organized clouds of mozzarella di bufala, ricotta, burrata, and latticini. Stainless-steel hotel pans are filled to overflowing with peperonata, garlic confit, fried rosemary, Sungold tomatoes, caramelized shallots, salsa romesco, and mustard vinaigrette, all waiting to be assembled with their corresponding cheese and served, front and center, on a sparkling white plate. The way Nancy preps these ingredients, sets her mise en place, and builds her antipasti seems effortless. How she flows from one action to the next, smiles kindly at the guests seated in front of her, and comes out from behind the bar to greet people are indelible lessons in the grace of hospitality.

My Amaro Bar is just to the right of Nancy's Mozz, and we spend our nights sharing intel about the most obscure aspects of our trade: the softness and durability of our beverage napkins,* the benefits of custom-made metal cocktail stirrers,† how the pars for burrata di Basilicata have to be doubled since the servers started a competition to see who could sell out fastest, which was happening before the first turn.

It's no secret that when I'm not at work, I'm rolling across L.A. with Sasha in search of a bar to buy. Every month he flies in and I pick him up at the airport where he arrives carrying a single rumpled plastic bag from his corner deli filled with toiletries and nothing else. We start off the same every time: 7&7s at King Eddy's and a Double-Double Animal Style from In-N-Out before following a meticulously mapped-out trajectory that keeps us off

---

* The standard black ones deteriorate in 2.5 seconds; ours last at least two drinks.

† Stainless-steel cocktail stirrers and straws can be washed and reused = sustainable practice.

the road between the blackout hours of 4 and 8 p.m., when every freeway and alternate route becomes a game of Chinese checkers. The archetypal stories about L.A. traffic, while anecdotally amusing, take up a large part of your brain once you've fallen prey to its horrors. It is important to remember to never take Sunset to Silver Lake when there's a Dodgers game. To never drive the 110 North to the 101 North from downtown's Fifth Street entrance because that interchange is *always* jammed. Sepulveda runs in a deceptively indirect diagonal from North Redondo (south L.A.) before hooking a surprise left at Culver City, which if you miss it, dumps you onto Jefferson which takes you into Baldwin Hills instead of say, Venice Beach.

Every new Angeleno will experience their *Falling Down** moment, when the driving becomes just . . . too . . . much. And they start screaming at their dashboard, brandishing their fists in the air and searching for an alternate route, except that's what *everyone else is doing*, so nothing changes and everyone gets home at the same time.

"AH YES. SUCCESS," SASHA SAYS, emerging from the airport carrying his usual rumpled deli bag and climbing into my car.

He holds out a few crumpled pieces of paper he's just extracted from the glove box.

"What's that?" I ask.

"My dry cleaning tickets. Hmm. There were three. I only have two here."

"I think two will be enough to get your clothes back," I assure him.

"Great. I will need to stop by first thing."

---

* A 1993 Michael Douglas/Joel Schumacher film.

Conversations about dry cleaning are quite common between the two of us. Sasha leaves shirts and pants at dry cleaners everywhere he travels so he doesn't get bogged down carrying anything aside from a phone, a copy of *The Economist*, and travel-size toiletries.

We crush our drinks, eat our burgers, hit the dry cleaner, and head toward Koreatown, along the edge of Skid Row—worse than any homeless situation I've seen in New York. In MacArthur Park, where the lake is rumored to be filled with bodies, stainless-steel Mexican fruit carts dot the sidewalks, their glass cases filled with mangos, pineapples, and watermelons sliced on the spot and tossed with chile, lime, and salt. Sign spinners* advertising deals on real estate or oil changes or grand openings entertain us as we idle at a stoplight on Coronado, and as we cross Virgil, all the signage switches from English and Spanish to Korean only.

I pull into a strip mall on Catalina—a nail salon, a hair-weaving shop, a dry cleaner (which Sasha notices immediately), and a shuttered Raffallo's Pizza. The bar we're here to see is tucked into the corner and has no sign. As we walk in, the transition from sun to level-seven gloom takes a moment to adjust to. The joint has no windows, no customers, and one backbar filled with unopened liquor bottles, which I take to mean they offer only bottle service.

Two Korean women in black, skintight dresses walk listlessly around the room like a scene from *Mulholland Drive*.† The minute

---

* "Human directionals" is what the outdoor advertising industry calls people who spin signs outside, since the signs are often shaped like arrows. There are various subspecies of the human directional, such as sign rockers (human directionals who move a sign from side to side), sign spinners (human directionals who spin, flip, and twirl signs), and sandwich men (human directionals who wear one sign on the front and one on the back and do not move the signs but stand inert).

† A very confusing David Lynch film.

we sit down on the cheap black aluminum stools at the faux-wood-paneled bar, one of them comes over and starts massaging Sasha's shoulders.

"Oh, shit," I lean toward him and say. "This is a hostess bar."

I laugh on the inside as Sasha melts with embarrassment—he's not only shy, but a true gentleman. I once had to drag him out of Jumbo's Clown Room because he kept handing $20 bills to every dancer who approached, hoping that plying them with money would convince them to find more suitable employment. Of course, this was not communicated and they kept dancing for him and he kept giving them more money.

Sasha's intentions were often in conflict with his actions.

Our conversation with the owner/landlord—an old Korean man with great hair and deep, grooved lines across his face I don't think are wrinkles—is futile, everything getting lost in translation, until his highly made-up wife who has to be twenty years younger than him slides the terms across the bar.

The deal points are bulleted on a clean white sheet of paper. No header or any indication of officiality in any way. There are seven items to address. I smile at the couple and huddle next to Sasha as we read through the document.

"Eric, we cannot have masseuses in our bar," Sasha whispers.

"You sure? You seemed to be enjoying your rub."

"It's strange, but oddly relaxing. Do I owe her money?"

"I think it's part of the sales pitch," I point to the barely legible xeroxed page in his hand. "These deal points look way out there. They want $80,000 in key money* which is insane because there's

---

* Monies you pay to buy the existing business or the "business opportunity," which consists of one or some combination of the following: liquor license, lease, FF&E (furniture, fixtures, and equipment), and intellectual property (name, concept, recipes).

no liquor license* included. The current market commands $55K to $80K for a license, which means we'll be at $160K of our budget and look around—"

We look around.

"We wouldn't want to keep any of the shit in here," I say.

"The stools could be worse," he says generously. "And with the lights kept low we could get away with it. However, the seat cushions would need to be changed to leather; pleather will make behinds sweat and guests will leave."

I give him the space to consider some of the more salient points.

"I like that there is a dry cleaners next door," he adds.

None of this is new to him, but it can be tough getting Sasha to focus. Sometimes he gets into the experience of things more than the actual business.

"Sash, what they're offering is a five-year lease with a five-year option to extend, and for all that up-front money it doesn't balance out—the rent is $2.90 per square foot for fifteen hundred square feet, which comes to forty cents more per foot than we wanted," I tell him patiently.

"Right," he says, bending forward. "It will take us three to four years just to realize our concept and cultivate our clientele and fingers crossed, we'll be paying off our debt in full by year four, just before having to exercise our option to sign another five-year lease." He pulls out the calculator on his phone and stabs at

---

* A liquor license can be obtained either via state lottery (luck of the draw) or more conveniently, through license negotiators. Market prices vary depending on where you're opening your spot and which type of license you require: a full liquor license, aka Type 48, which does not require food sales and is usually the most attractive, or Type 47, which requires an establishment to be considered a bona fide eating place with a full kitchen that either sells a certain ratio of food to liquor sales or has an operating-hours requirement for food service or a combination of the two.

numbers. "If our rent is $4,350 per month, which ideally shouldn't exceed 7 percent of our monthly sales, that means we will need to make, at the very least, $62,000 per month—multiply that by 12 and then divide by 52, which is roughly $14,300 per week—and with $12 drinks that's about 170 cocktails per night." He pauses and bends even further forward, almost speaking into his hands. "We want a much longer lease. Like a ten-year with a five-year option or a ten-year option. What is our budget by the way?"

"We agreed to $350K or less," I remind him, and to me, even that sounds high, never having spent six figures on anything in my life.

In the end, the deal-breaker is the bullet point on the NNN* where this mafioso-exuding landlord/owner has CAM charges set to start at twenty-five cents per square foot, with no cap.† And while the decrepit strip mall might be considered an appropriately "L.A. location" for a classy classic cocktail bar, I have a feeling the landlord thinks he can gouge us.

I peel the hostess massaging Sasha's shoulders away and get us the fuck out of there, heading straight down the block to the fifty-six-year-old nautical-themed HMS Bounty—a bar/restaurant I

---

* NNN is an abbreviation for the three (or triple) nets found in commercial leases: property tax, property insurance, and CAM (common area maintenance) charges. All three are passed on to the lessee in addition to the base rent. Gross leases also exist, which include all of the above in the base rent price, but the majority of leases out there are NNN.

† Through the initial lease term and subsequent options, you want to negotiate a market cost per square foot and cap the annual rent increases at a fixed percentage (3 percent is common) or at a fair CPI (consumer price index) for the local market. It's also in a tenant's best interest to have a cap on any increases in CAM charges. A knowledgeable landlord will agree to caps only on controllable items, like the salary for the building's maintenance person, which all the tenants contribute to. Uncontrollable items, like utilities and trash removal (which in L.A. is now dictated by the city), are services provided by third-party vendors that the landlord has no control over, so caps on these costs are unlikely.

would saw my right arm off to own, which seems like a hidden caveat some slumlord might include as a clause in a lease agreement.

Our search continues.

SASHA AND I DON'T JELL with the suburban vibe of North Hollywood or Burbank, but up off Lankershim Boulevard there's a bunch of acting studios and theaters, and if there's one thing I know, actors drink. We get a tip that a hole-in-the-wall off Magnolia might be for sale and walk in on a Wednesday at noon. The door takes a shoulder shove to open; the air reeks of dank stale beer. A television rests on top of a defunct vending machine completely empty save for a deflated bag of Lay's potato chips. An old woman stares in rapture at QVC on the TV as a post office employee, fully suited, bag of mail on the floor, swills a beer and the bartender curses at him for being fat and lazy. We ask for Bill and Judith.

"Bill! They're here!" the bartender yells over her shoulder while giving us the evil eye.

We ask for the P&Ls* so we can confirm the annual $500K gross sales they claim, but when we receive them, half of the line items are redacted. Not very transparent. They claim they're netting 30 percent, but without being able to comb their financials it's hard to believe, especially in the bar's run-down state. Upon further investigation we learn they have a violation looming on their CUP.†

We bounce.

---

* Profit and loss statements.

† A conditional use permit determines what special conditions you need to run your joint, e.g., amount of parking, hours of operation, permissible gaming, live entertainment, security requirements, etc.

⊷═◉ ◉═⊶

IN SANTA MONICA THE STOREFRONTS lack historical character, and rents are astronomical at $6 or more a square foot, which makes no sense for a craft cocktail bar running margins just a touch over a restaurant's bottom line. Further south toward Inglewood we find a brilliant jewel of a spot, which in its R&B heyday must have had Cadillacs lining up in front. We can envision returning the place to its former shine, but the neighborhood doesn't look like it can support it, so it's on to Manhattan Beach which is too surf brah for us, though its proximity to LAX is a plus since Sasha is prone to missing his flights.

The way we're searching for our dream bar isn't working. I know we need to change up our tactics, which is something I'm all too familiar with. I've always learned things in messed-up ways. As a child with dyslexia, I did everything backwards. Literally. In fourth grade, Mrs. Pillsbury* thought I had some marbles missing. Thankfully my parents understood I wasn't a moron and provided the resources necessary for me to conquer a different brain—not a defective one, just one that chose the path less traveled.

We decide to look for places that aren't advertised for sale and hit every Mexican beer bar in Boyle Heights, Lincoln Heights, and Cypress Park, places with worn linoleum floors and scarred laminate bar tops, run by *abuelas* who only serve *cervezas*. Later, we write letters to the owners to petition them to sell, and when we don't hear back, rewrite them in Spanish.† When we do hear back, the terms they want are unrealistic, and when they aren't and we

---

* Real name.

† And by "we" I mean Hector, my barback at Osteria Mozza, translates them in exchange for bottles of mezcal.

contact ABC,* we're told their Type 40 license (which allows only beer) can't be changed to a 47 or 48 (full liquor) because there's a church, hospital, school, or nonprofit youth facility nearby.†

I don't know what I expected this process to be like, but driving around L.A. and looking at space after space that's too big or not licensable or has liens on the business or ones that are so fucked up they'd needed tons of money sunk into them for any concept short of postapocalyptic is debilitating. Asking the same questions over and over again is like *Waiting for Godot*: *How long a lease can we get? Do we have first refusal to buy the property if the owner decides to sell?‡ What are the per-square-foot terms? Are LADWP connections and meters intact and up to code?§ What about the "good guy" clause?¶ What about our pro forma** and business plan? What does the Thomas Guide†† say is the best route from Tujunga to Topanga?*

The responses vary; the result stays the same.

We grow road weary but never tire of the city's landmarks: the circus-bulb lights of the Santa Monica Ferris Wheel, the sculptural spires of Watts Towers, the high, white dome of the Griffith Observatory, the absurdist Randy's Donuts with its giant doughnut on

---

* Alcohol Beverage Control.

† A retail space that sells liquor must be at least six hundred feet away from the aforementioned.

‡ Make sure you do, so you can be your own landlord!

§ Los Angeles Department of Water and Power: sewer, gas, water lines, meters, electrical poles with 400-amp three-phase power. You want to ensure all of these "stub-ins" connect to your property.

¶ A good guy clause limits the tenant's liability if a lease is terminated early. An example of this is if after securing a lease your ABC license or license transfer doesn't come through, it annuls the lease agreement.

** Latin for "as a matter of form." In business, it's a tool used to map the financial projections of a business over the course of five to ten years.

†† A binder of maps; the precursor to navigation apps.

the roof, the ocean, the mountains, the windy roads bordered by Seussian palms, and the forty-five-foot Hollywood sign—a literal reminder of where we are and a metaphoric reminder of where I want to be.

IT TAKES SIX MONTHS OF pounding the pavement until a combination of luck, destiny, and timing brings the rainmaker of DTLA* bar culture to Mozza when I'm working. Cedd Moses—barfly entrepreneur who owns the Golden Gopher, Broadway Bar, and Seven Grand, and son of the late, renowned abstract "Cool School" painter Ed Moses—pulls up a barstool and regales me with stories of the burgeoning cocktail scene downtown. His management company, 213 Hospitality, has been championing the neighborhood's *Escape from L.A.* vibe, turning defunct historic spaces into dark, sexy places to drink. To Cedd, dirty streets, abandoned corridors, and crumbling art deco buildings that block out the sun are the best parts of L.A., and when he hears I come from the Milk & Honey family, and that Sasha and I are looking for a bar, he thinks we may be of similar minds.

Two weeks after our auspicious encounter, Cedd invites me to join him and six other L.A. bar hounds† for a weekend of R&D in San Francisco, which includes but is not limited to beet-root-infused drinks at Absinthe, mai tais at the Slanted Door, crafty cocktails at Bourbon & Branch, and late-night falling-asleep-in-my-burger burgers at Nopa. As we drive back to SoCal—heads throbbing, Tylenols popping—there's a palpable sense of excitement in the

---

* Downtown Los Angeles.

† Vincenzo Marianella, Aidan Demarest, Damian Windsor, Marcos Tello, Patrick Kelly, and Ricki Kline. All dudes. This male-to-female ratio imbalance has, thankfully, changed.

sleep-deprived air. We don't quite know what we're doing, or why, or what any of it means, but it feels exciting. In the movie version of the moment, we would all smile slyly at one another in silent acknowledgment that each of us is about to set the city of Los Angeles on fire, but in real life we're too hungover to do anything beyond suck up all the available electrolytes and keep whoever's driving awake by blasting Zeppelin.

The next time Sasha's in L.A., he, Cedd, and I eat bibimbap* at Soot Bull Jeep before Cedd shows us a space in the back of Cole's French Dip—a restaurant that's been in operation since 1908 which he's recently taken possession of.

It's a storage room, and it's dark and windowless and looks, well, like a dark windowless storage room. But for two New Yorkers looking for a home, it glitters like a jewel, and, several emails and phone calls later, the three of us agree to partner up.

After all these miles, it happens in a New York minute.

---

* "Mixed rice" in Korean, bibimbap traditionally consists of a bowl of warm white rice topped with namul (seasoned vegetables), gochujang (chile pepper paste), a cooked egg, and sliced meat.

CHAPTER THREE

BUILDING THE BAR

IMPOSSIBLE N'EST PAS FRANÇAIS.[*]

—*Viviane Alperin, my maman, quoting Napoleon Bonaparte*

Call me naive, but I thought once we had the space secured, I'd be knee-deep in renovations. I'm sure that's hilarious to anyone who's been down this road before, but I haven't. I'm learning everything for the first time, by the seat of my pants. While Sasha's a great wealth of knowledge about a lot of things, he's not particularly *practical*. When I turn to him for advice, I'm reminded that even with three bars under his belt he's never built out in a traditional manner, opening M&H on a credit card with a $10K limit and then piece by piece, with what he could pull from the weekly till, trying to finish what he started.

Staff who worked there will tell you that for the most part, nothing was ever "done."

---

[*] "Impossible is not French."

I made one irreversible mistake three months ago when I quit Mozza to focus on our project, assuming we'd be up and running and making money in no time. So I burned through what little savings I had and now spend my days building the bar and my nights working at the Doheny—Cedd's new private members' club on Olympic at Figueroa, named after E. L. Doheny, an oil tycoon who launched the petroleum boom in SoCal by drilling its first oil well and then constructed a building in his own name. Our bar is accessed through the garage and lives inside the ground-floor greenhouse, off the lobby toward the back. The concept is the finest bartenders making the finest cocktails serving the finest spirits, accompanied by bar bites of crudités and caviar from Grace and BLD's Neal Fraser. The mirrors are etched by Shepard Fairey. The membership fee is $5,000 annually and doesn't include a bar tab. No one . . . really . . . gets it . . . which means it's empty enough that GM Aidan Demarest, assistant manager Kate Grutman, and I don't have much to do, so we entertain ourselves with booze and blow and naps in the banquettes, naked theatrics and wardrobe changes in the office dungeon—the most CO_2-poisoned room you ever did see. When we want to hone our Champagne-opening skills, we grab a magnum of Perrier-Jouët and saber off the top with a bread knife. I'm destroying my brain and body working and playing 24/7, all of which is unsurprisingly wreaking havoc on my love life. Courtney is unconditionally supportive of the bar project, and I'm unconditionally supportive of her acting career, and if they both succeed it'll be great! But right now, we're ships in the night.

"Hey," I whisper, crawling in the door at 6 a.m. from an exceptionally mind-numbing night at the Doheny.

"Hey," Courtney says, in a normal tone of voice because she's just woken up and is on her way to teach yoga. She hands me a glass of green juice she made in an attempt to keep me alive, and it makes me want to cry because I want to be the kind of guy who drinks green juice but I'm not. Not right now anyway. Right now I'm the kind of guy who polishes off half a bottle of mezcal a night, consumes a gram of cocaine, and winds up in situations that push the boundaries of harmless flirtation.

"I'm gonna down this after a long, hot shower," I say, holding up the glass.

"All right." She hefts her bag over her shoulder. "I hope you get some sleep."

"Yeah, no," I shake my head. "I gotta be in in two hours."

"Well—" She smiles and kisses me and I wonder if there's time for—

"I'm running late," she says. And I know it's true because she's always running late, but man what I would give for a quickie . . .

THREE HOURS LATER, I WALK into our construction site to find workers laying down sheets of penny tile without using plastic spacers, which means that every seam is either too far from the other line of tile or jammed together. Granted, I snorted the last of my baggie to kick-start the day, but I don't think I'm seeing things.

"It'll be dark in here," the tile guy tells me when I point it out. "No one will notice."

"I notice," I say. "You have to fix it."

"But we're halfway done."

This doesn't need to be my fight right now, so I call our general contractor; we'll call him Paul Croft.

The next morning, Paul and I stand together staring at the floor.

"What?" he asks, looking at the wonky tiles.

"What do you mean, *what*?"

"So they're off a little."

"*And* don't follow the pattern."

"What pattern?"

I'm about to lose it.

"Well they can pop out the white tiles and put black ones in for that," Paul says before I can blow. "Besides, it'll be dark in here. No one will see."

"I will see!" I yell. "I will feel it under my feet and others will too. Even if only one person a night notices, that is too many. These details matter. This is why we have meetings and measure shit and agree on things!" I'm really close to his face and he's six foot four, which makes it really easy for him to grab my shoulders and shake me while yelling: "I haven't paid my mortgage in two months you little shit! I don't see my kid anymore because I'm running too many jobs so don't tell me what to do!"

Since this kind of interaction with a guy who could squash me is a sensible thing to walk away from, I grab his shoulders, and now we're shaking each other, or he's really just shaking me and it looks like I'm shaking him but it's because I'm hanging on, when two arms grab me from behind and another two come in between me and Paul and pry us apart.

The chef from Cole's and his prep guy.

We're all breathing heavy. I see tears in Paul's eyes.

"Paul man, I'm sorry," I tell him. "Real-life stuff with home and family is tough. I feel you. But in all fairness that isn't my concern. I need these tiles done right, and you know that's what's fair."

The next morning the tile subs re-lay the penny tile. This time with spacers. Which is good news. The bad news is that my designer, Ricki Kline, who looks like what would happen if Hemingway did a stint as a Hells Angel and then decided to get into design, tells me he doesn't have the drawings I need for the furniture-and-fixtures layout.

"Fuck you!" I scream at him, because this has happened before in the same general pattern: I ask for the drawings so we can decide on the look and materials and then contract a mill worker, lighting designer, and stainless-steel fabricator; Ricki says he doesn't have the drawings yet; I ask when he's going to have them; he tells me it's all going to be fine, I should chill out; and I lose my shit.

It's not that Ricki and I dislike each other, it's that the way we work is wildly different. Maybe it's an East Coast versus West Coast thing, and while I feel the need to be in constant motion, Ricki seems happy to wait and see. I know he's gone through this process a dozen times and that makes him the pro, but this is my maiden voyage, and every single thing that happens (or doesn't) feels cataclysmic. And while no individual detail of this endeavor is particularly difficult, because I can't actually *do anything*, all my thoughts and ideas are getting turned into lists miles long, and the longer my lists grow, the more anxious I feel that this was all a mistake. That everything is going to fall apart, and I can't afford to entertain that thought because I'm close to losing my mind, and just as I'm about to apologize to Ricki—not because I think I did anything wrong but just to defuse the situation—his phone rings and he leaves the bar, which is Sasha's cue to walk in through the back wearing his pocket watch, which doesn't work but at least reminds him that time exists.

"BUILDING AND OPENING A BAR is always a moving target, and it's usually moving away from you," Cedd tells us over French dip at Cole's. I take copious longhand notes as he talks and Sasha synthesizes Cedd's knowledge with neither pen nor paper.

"I get it, the process of working with Ricki can be frustrating," he continues. "But the result is really good, and he's affordable. You usually only get two out of three when you work with a vendor— good, cheap, or fast. And while Ricki's not fast, the price is right, and at the end of the day, his bars look great."

We're still in the getting-to-know-you phase, and while I know Cedd is our partner, I've been hesitant to ask him for advice. Not because he's intimidating, which is what a lot of people think (he's just super tall and pretty weird), but because I can be a little arrogant. I think I can do everything on my own until I realize I can't do everything on my own, and then I ask for help. It's not a great trait, and something I'm actively working on. I try to remember that asking for help doesn't mean I'm incapable, just that I can't do everything alone. No one can. I know that. I have to remember that.

"While we wait on the plans, we can work on the concept and details for build-out," Cedd says as my French dip disintegrates in a hypertension salt mine of au jus.

"Go get bids for the stuff you already know you want. Let's make what we can offsite, and once we get the green light we'll be ready to rock."

SUNNY'S FIXTURES COMES IN WITH the best bid for bar equipment, which I mention to one of our suppliers in passing.

"Oh yeah, you should go see Sunny in person," he snickers.

"What? Why?"

"I bet you can get him to go even lower," he says, wiggling his eyebrows. "If you know what I mean."

I do not know what he means, but our budget is busting at the seams and every little bit helps, so I drive over to Venice and Western, one of L.A.'s more depressing no-man's-lands: AutoZone auto parts. 7 Dias Tire. S&J Wilshire Tow. Food 4 Less. The Korean Evangelical Zion Church, Triangular Church of Religious Sciences, Congregational Church of Christian Fellowship, and Young Saeng Presbyterian Church battle for souls on either side of the median. And then Sunny's Fixtures—a low-lying stucco building that houses a huge depot of restaurant and bar supply items.

Inside, the place swarms with employees stocking shelves, uncrating bulk orders, and ringing up customers. I ask a cashier if I can speak to Sunny, reference the bid, and am told it will be ten minutes. After thirty, having walked up and down most of the aisles getting ideas and taking stock, I approach the cashier again.

"Is Sunny available?" I ask, and for some reason, this time am immediately escorted to a back office where a tall, stocky Latin man swivels toward me in his desk chair.

"Hello Sunny," I say.

"He not Sunny! I'm Sunny!"

I flip around to see the skinniest, shortest, most effeminate Vietnamese man I have ever met.

"You Eric. You cute! Whatchyou want?"

I think maybe I'm being punked, but when I look around the office, everyone appears unperturbed.

"Ah, Sunny. Hi. Yes. I'm Eric. I have the bid you sent me and would like to go over it in hopes of some price adjustments."

"You no like price? Too bad. Whatchyou do for me? Huh?"

"Uhh . . ."

"I kidding Eric! You no laugh. Oy, white boys. So cute. Okay come here. We talk."

After an hour of haggling and deflecting Sunny's flirtations without seeming rude, he lowers his prices when I throw in a night of free drinks once the bar opens.

"See you at the no-name bar Eric!" He waves as I leave. For some reason he thinks it's hilarious we don't yet have a name. "See you there!"

⤞⟹ ⟸⤝

"HEY JOE," I SAY WHEN my dad answers the phone. I call him Joe because that's his name and it's how I've always addressed him. Joe is also a lawyer and, specifically, my lawyer; everyone should have a bit of nepotism to cash in on.

"Eric!" he says at the same time as my maman says, *"Eric."* I believe they have their own phones, but they're always speaking on one at the same time. My family are also FaceTimers.

"Hey are you . . . you guys are . . ." Their heads move in and out of the frame—there's the ceiling, then the floor, then back to their faces, and all the while they're telling me about their day. I love my parents, but we just talked yesterday and there seems to be a disproportionate amount of news to relay.

When the phone settles and we're caught up on all that's transpired in the past twenty-four hours, I ask Joe if he has time to help me with the partner contract I've been putting off dealing with. I know it's irrational, but thinking through all the ways shit can go sideways seems like getting divorced before getting married.

"Which is why they created prenups," Joe argued when I used this as an excuse for not having a contract in place. "The odds are not in anyone's favor that everything's going to work out—in love or in business—so you have to protect yourself the best you can. Think of it like a map that expresses what you want and what you expect. It takes the guesswork out of who's in charge of what, for how long, and why."

"The most salient points to include are percentage of ownership," Joe tells me now, diving right in.

I grab the pad of paper I got from a sponsored liquor event (my loft is littered with swag) and scribble, "percentage of ownership."

"It means who owns how much of the pie."

"Right, right," I nod, picturing a pie chart from grade school that showed the important food groups, which back in the eighties included evaporated milk, canned vegetables, and fortified margarine.

"Have you guys discussed creative control?"

"I mean it's, you know . . . we're figuring it out."

"Okay. Great," says Joe. "Let's say you have an idea about curtains—"

"Curtains?"

"Yes. Eric. Let's say there's a *design element* you like, but Sasha doesn't like it, and another guy doesn't like it, and someone else *does* like it. Who has creative control, how is it brokered, and who has say over what? It can be different for every aspect of the bar; for instance, maybe you have final say over design and Sasha has final say over product, and what's the money guy's name—"

"Cedd," I remind him.

"Maybe Cedd gets final say on the name."

I gotta get on the name.

"You need to calculate sweat equity versus capital interest," he continues, faster than I can write.

"That means how much I'm going to work because I'm not putting in any money?"

"Right," Joe says. He stands up, sets down the phone, and starts to pace, so now my visual is the living room and every once in a while, the lower half of his body passing by. "You have to calculate how the percentage of ownership a partner puts in translates to your work hours."

"How do I do that?" I ask.

"It's up to you. Whatever you want really, you just have to agree."

I write, "agree to work=$$$."

"I assume you're going to be the one dealing with the day-to-day?" he asks.

"Yeah."

"Well you still need to put that down on paper and make it part of your management agreement. Have you guys aligned on a business concept?"

I mentally shuffle through the five thousand conversations the three of us have had. "Basically?"

"Put it into words Eric and *write it down*," Joe admonishes, leaning in *so close* to the camera. "The contract needs to state that there's one idea, not seven, and then each of you sign off on it and no one can come back and say, 'Hey, I thought we were opening a French bakery.'"

"All right, all right, I got it," I tell him and write, "One idea not seven. French bakery."

Joe sits back down.

"Your voting rights and minority member protections should be based on percentage of ownership and/or potential issues in question."

To which I ask, "What?"

"Eric," he leans in *so close again*. "There will be times where the partners need to decide on allocation of monies for profit distributions or to make a capital improvement to the bar, so you want to make sure your vote or opinion matters and holds weight in the decision-making process. You also want to make sure that you exercise creative control over the service plan that you and Sasha hang your hats on, which isn't to say you aren't going to listen to your partners, but, like I said before, in the end, someone has to have the final say if there is a disagreement. Hopefully that doesn't happen, and you sign the contract, put it in a drawer, and deal with each other like friends and professionals. But in the end, you never know, so you have to hash it all out now."

"Okay," I nod, my head spinning.

"And noncompete clauses?"

Joe shakes his head at my blank face.

"Eric—"

"I know! Noncompete. Who each party can and can't work with and how close someone can open a competing bar to the business we've opened together in the future."

In the background, the sound of my mother setting dishes down on the dining room table slams me with homesickness.

"What's for dinner?" I ask.

"Hey Viviane!" Joe yells. *"What's for dinner?"*

I can't hear what she says—she's too far from the phone—but my dad turns back to me and says, "Couscous chicken tagine. Your favorite."

WHEN IT COMES TIME TO build the actual bar part of the bar, Ricki and I, no surprise, get in a fight. While he has an eye for a bar's aesthetics, he's never actually *worked in a bar* and knows nothing about what's necessary. Chefs have been designing their workflow in kitchens for decades, I argue. Why shouldn't we? He yells. I yell. He eventually, and begrudgingly, defers.

The bar we're building will have two stations that basically mirror each other: the service well, which is where drinks for the seated guests are built, and the personality well, which handles standing-room guests and overflow tickets from service. There'll be a thirty-six-inch Perlick glass froster for chilled glassware and block ice for cocktails. Next to that, a twenty-four-inch double sink for rinsing and dumping. This is also where the bartenders will stand, pulling bottles from a single rail in front of them. Next is a twelve-inch crushed-ice bin for juleps, fixes, chilling bar tools, and a shared Perlick bottle cooler that holds beer, wine, vermouth, and soda.

Our work space for building drinks is an eighteen-inch-deep

stainless-steel grated bar top. It's a large scupper,* which might seem like overkill but makes for a virtually spill-free zone, saving bartenders and guests alike. Also, it's what I worked on at Little Branch and there's no reason to reinvent the wheel.

Ricki tries to get me to back down on using stainless steel for wall flashing,† but I won't budge because fiber-reinforced plastic, or FRP—a composite material made of a polymer matrix that usually comes in white textured panels—looks cheap and yellows over time. Gross.

The industry standard heights of prefab sinks and ice bins are too low for my taste, so I fight to get custom legs installed to lift them from twenty-nine inches to thirty-two and a half. This way the bartenders won't have to hunch over to work, which is not only hell on their backs but just looks bad.

I got the under-bar equipment as narrow as possible—about nineteen inches deep—which allows the bartender to be closer to the guest and offer engaging and upright service. If I had the budget to build custom stainless equipment, I would have gone even narrower.

The bar top itself is forty-two inches off the ground to the top side and thirty inches wide from bartender to guest. It sucks to have to yell or cup our ears because we can't hear a word anyone's saying or when we have to reach over really far to serve a drink. Additionally, the scupper extends twelve inches on the bartender side over the sinks and ice bin, allowing for a cozier workspace.

---

* The amount of space provided for the bartender to make drinks. Some are just four-inch rails with bar mats, but I always make mine out of stainless steel at a minimum of eight inches deep, with grates that allow drainage.

† Thin pieces of material installed to prevent the passage of water into a structure from a joint or as part of a weather-resistant barrier system. Bars are like boats—very much a maritime environment. Stainless all the way! Also, it's easier to clean and looks pro.

The backbar shelves are four inches deep and stacked to have a minimum of fourteen inches between them, which will accommodate the width and height of most bottles on the market. Except Galliano, but who the fuck drinks Harvey Wallbangers?[*]

I completely screw the pooch with our booths and tables. It isn't until they're bolted into place that I realize there's too much space between the backrests and table edges, which means when guests recline, they're going to be too far away from those seated across from them. This might fly in a diner but is no good for intimate cocktail conversation.

Bigger isn't always better. Lesson learned.

My daily *Goodfellas* commute from my loft to the bar means I walk out my front door, walk down the hall to the service elevator, walk through the parking garage into Cole's, cross their kitchen, and end up in our work site/closet. I then reverse my steps to get back home with a slight variation: I walk out the back of the work site into the dimly lit hallway that ends in the parking garage where I take the service elevator up to the seventh floor and home sweet home! Which is to say I never see the outdoors. I like to pretend I'm on a spaceship and if I step outside I will perish. I need this kind of incentive in order to spend my days sitting in front of my computer setting up our LADWP, credit card processing, payroll, trash, internet, and phone service accounts, filling out forms for our workers' comp and general liability insurance, claiming our social media accounts, writing and printing a training manual, scheduling staff, doing inventory, creating iPod playlists which is

---

* This drink is vodka, orange juice, and a float of Galliano—a golden yellow sweet herbal liqueur from Italy that comes in a ridiculously tall bottle resembling a conical skyscraper. Invented by Donato "Duke" Antone at his L.A. bar Blackwatch for a surfer named Tom Harvey, who got so lit one night on this drink, he started running into walls. Or so we all thought, until Robert Simonson debunked the theory, stating that "no sane person ever believed that story."

shockingly time-consuming: arranging songs with similar tempos into folders and curating seven hours of music for seven nights a week with a little variety so no one goes insane. That's forty-nine hours of music and I still have to design drink tokens, matches, business cards, and cocktail menus—oh shit, the menus . . .

"DON'T WALK LIKE A BITCH!" a pantless homeless woman on a bicycle screams at me as I cross Main.

My first encounter with the outside world in months.

Joke-Man stops me as I cross Sixth. "What's the best nation in the world?" he asks.

"What's that buddy?" I say.

"*Donation*," he grins.

I hand him a dollar.

Woo! Human interaction. Vitamin D!

I walk with my arms swinging like a musical through downtown to Moskatels to gather menu-making supplies. Moskatels is the holy grail for set designers, florists, Halloween shoppers, and bar and restaurant owners looking for random shit. It's located smack-dab in between the flower market and Santee Alley, which is an actual alley filled with tiny shops and kiosks packed with piñatas, fake designer bags, costume jewelry, and electronics. The smell of lengua tacos from the dozens of vendors packed into the alley is so strong it nearly knocks me off my feet.

Moskatels overwhelms me. Feathers and paint and fake flowers at every turn, but I eventually find what I need—cardboard, yarn, paper—and rush back home. Dump out my purchases. Print out the image of a cocktail coupe my maman designed based on an old Cole's diner mat and glue it to a piece of cardboard. Flip the cardboard over and cover it with a printout of the cocktail list. Hole-punch all four corners and thread them with yarn. Think the thing would look good in wood. Realize I'd need a branding iron for the coupe. Google "branding irons." Find "brandingirons.com." Get stoned. Eat three Milano Double Dark Chocolate cookies and chug a glass of water. The next day, cardboard turns to wood. A day after that, yarn to leather. And the day after that, a branding iron with the shape of a coupe arrives in the mail. I grab it, grab my menu mock-up, and head down to the garage. It's before 4 and after noon, so it's only forty-five minutes before I'm pulling into a barren lot on San Fernando in Mount Washington and parking next to a Ford F-150. The SoCal sun bounces off my pale skin as I grab my home-ec menu idea, and quickly make my way across the lot toward a heavy steel door.

Inside is a five-thousand-square-foot windowless warehouse, dimly lit by makeshift lights slung over support beams. I'm met by the whirring sound of a table saw blade ripping down planks of wood. Table saws are fucking frightening and will take fingers off like it's their job. I have a healthy respect for them because my buddy Ben got part of his finger nicked off in one and I had to race him to the ER, blood spurting through piles of shop rags. Another time I got assaulted by a plank of wood I tried to run through a table saw which shot back into my solar plexus and knocked the wind out of me.

"Yo! I got your message," my friend and master carpenter Eric Thorne calls out as he stops the lumber-eating monster. I can now hear Johnny Cash playing low on a dusty one-speaker radio. "Carving menus into wood feels like a shit ton of work," he says, grabbing a broom to sweep up the sawdust.

"Oh, of course, no," I say, and hold out my mock-up. "I was thinking we could make boards and lace menus onto them."

He stops his incessant sweeping for thirty seconds, looks at my mess of cardboard and yarn, and says, "I got an idea."

We spend the rest of the day cutting, sanding, branding, and varnishing scrap mahogany Thorne pulls from various places around the warehouse—the floor, the bench, hundreds of shelves—and by the end of the day we have forty-two menu boards.

"You Thorned it!" I say, toasting him with a beer. It's the highest compliment among our friends, because Thorne's a true MacGyver.

"*We Thorned it*," he tells me, and I'm a sensitive guy so this makes me tear up.

EVERY INSPECTION IS DONE EXCEPT for health, which we wait for. And wait for. And wait a little longer for. One day I'm waiting at Cole's bar eating a Caesar with roasted turkey and notice a soft-spoken walrus of a man leaning into the rail in order to be heard better. He's asking the bartender for an Aviation, but the bartender is new and not yet versed in all the classics, so I chime in and tell him the drink he wants will be available when we open in back.

"Oh," he nods, his long red hair sweeping his shoulder. "End of February?"

"As long as we pass inspections," I laugh-ish.

He pauses and smiles. "Are you Eric?"

"I am."

"I'm Jonathan," he says. "I believe we have an interview scheduled . . ."

This is how I meet Jonathan Gold, the then–*LA Weekly* food and beverage critic. J.G. likes classic cocktails *a lot* and regales me with things he's memorized from the old cocktail books that line the stairway of his home in Pasadena. He always says, "But I'm sure you know that already," after describing an arcane recipe or long-forgotten ingredient.

Jonathan is brilliant, funny, quirky, and well-read, wears band

T-shirts, and only wants to celebrate people, the more unknown the better. Soon after we open, he'll write a cover piece for the *Weekly* about L.A.'s burgeoning classic cocktail scene* and in so doing, change the landscape of the city's drinking culture forever.

"What are you calling it?" Jonathan asks before we part.

"The bar?" I ask.

He nods.

We still don't have a name.

"HERE—" JOAN, OUR PR REP, hands me and Sasha a VHS cassette coated with a thin layer of airborne kitchen grease and the title *Steel Rails: Private Varnish* on the cover. "This might give you some ideas."

Steel Rails: Private Varnish features eighty-seven minutes of glory shots of private varnish train cars around the United States. Watching it, Sasha and I learn that the term "varnish" is a reference to the high sheen on the wood of the private trollies which the titans and business magnates back in the day used as their own commuters. When standing on the train platform above Cole's, you would see one of these private railroad cars coming down the tracks and call out, "Here comes the varnish!"

We say it out loud a few times. "The Varnish. The Varnish! *The Varnish.*" Think about our trolley-inspired booths. The history of the Pacific Electric. Write "The Varnish" down in ink. Type it out in different fonts. Laugh at how guests are going to *get varnished at The Varnish.*

Then we get word a health inspector is coming.

---

* The article, titled "The New Cocktailians," was published in March 2009 and was nominated for a James Beard Award for journalism.

I glance around the bar which is still a construction war zone—piles of tile, paint cans, bits of wood, gobs of electrical conduit, glassware I preordered, which I'll need to hide because you can't stock a damn thing until you pass.

"When the health department comes through, you want to look like you're ready to open," Cedd tells me. So I call Paul and tell him to clean his shit up and then call Manuel from Smart Clean to give the bar a once-over.

"You also need to make sure you've considered everything you could possibly think of, and you know, sometimes they find stuff you didn't even know you needed, like Cole's failed its first health inspection because we didn't have a grease trap. We had to push back our opening date and spend a fortune building a giant underground grease interceptor that looked like a submarine, which is ironic because when we bought the place they didn't even *have* a grease trap. Asbestos was peeling from the kitchen ceiling. There was standing urine in the kitchen. You need to know how to use all the equipment."

"Right," I say, trying to parse the information. Cedd speaks in long, meandering paragraphs with nonlinear thought patterns. "Standing urine. Learn to use the equipment. Got it."

But days go by and no inspector shows. I feel like Ricki told me the wrong date, so I call our expeditor, Fast Eddie, and he confirms we're scheduled between 1 and 7 p.m. on Thursday, February 5.

Tomorrow.

I breathe a sigh of relief, having projected February 24 for opening night because it's my dad's birthday, which feels like a good omen, and Joan needed a date to give media outlets to make sure they save a slot for us in their publication.

The next afternoon I plant myself at a booth to wait. I feel like I'm in seventh period in high school waiting for the final bell and the big hand keeps clicking backwards. An hour passes. I've been sitting in silence. It's oddly therapeutic. It's been nonstop for months now and it all feels like it's hinging on this moment and there's nothing more I can do.

"ERIC ALPERIN?" MY HEAD SHOOTS up off the table. I'm not sure when it happened, but I've fallen asleep.

"Huh? Yes! Hi! That's me!"

"Greg Woods from health." He walks toward me, hand extended. "You look about right for opening a bar. Tiring stuff."

"Oh yeah, you're not wrong," I say. "I don't even remember falling asleep. Good to meet you. So how do we do this? I'll be honest, it's my first time."

Greg turns out to be a really nice guy, not the hard-nosed inspector I was warned about. He runs the water in a sink to make sure it's hot throughout the bar line, checks that the fridge and freezer are the right temperatures—41 degrees or below for all refrigeration—makes sure there are casters, aka wheels, on all the equipment so it can be moved easily for cleaning, checks to see that the space between the backbar and edge of the bar equipment is at least thirty-six inches—an ADA* requirement for any handicapped bartenders—and watches me work the goddamn Hobart glasswasher, which I don't know how to use but learn on the fly, and get it up to 180 degrees, the temperature needed for sanitizing glasses.

---

* The Americans with Disabilities Act is a civil rights law that prohibits discrimination against individuals with disabilities in any area of public life, including jobs, schools, transportation, and all public and private places open to the general public. If you don't bring all areas of your establishment up to code during initial construction, it can become a super expensive liability later on with potential drive-by lawsuits, fines, fees, consultants, and repairs.

"Well," Greg says, leafing through the papers on his clipboard. "Not everything is in order."

"Wait, what?" I ask, a crackle in my voice because he was so nice and never said anything about anything being messed up and now I'm wondering if it was all an act.

"You didn't install casters on your backbar fridge, and even though it's meant to look like it's a part of the custom stainless, it's still not physically attached to the backbar unit—" He pauses as if expecting me to finish his sentence, and when I don't he continues, "Which means random stuff can get underneath there which you will need to clean out."

"Oh shit," I say. "I promise to pull it out weekly. Really."

"I believe you Eric, but it's not about me or you. It's about the future and who might be working or replacing you."

"Oh man. I hear you," I sigh, crestfallen, feeling like this is going to earn us a delay.

"But I'll tell you what," Greg scribbles a number on a piece of paper. "Here's my cell. Get those casters on by Monday, snap a pic, and text it to me. I'll pass you today, but I expect that photo ASAP."

THERE'S NO TIME TO CELEBRATE this heart-stopping milestone because I still have to finalize our press release, schedule a photo shoot in the bar with cocktails, order booze, bring back the glassware from hiding, call in staff to stock the bar, and schedule a friends-and-family night, and even though I've already chosen what I want on my backbar, I find myself entertaining shit tons of ambassadors and liquor reps snake-oiling their booze, coming in hot and heavy and pushing vodkas especially. I keep saying, "Thank you, but no." One afternoon I tell a Southern Wine & Spirits rep I plan to carry only two vodkas. He chuckles and says, "Oh

Eric, this is L.A. We'll break you down and you'll have the Absoluts and Grey Gooses up there in no time."

I thank him kindly for his time and call the Southern help desk to request a new rep.

I'm on overdrive, full metal jacket, so when my buddy Jeff Hollinger* from Absinthe in San Francisco calls and asks how I'm doing, I ramble on and on about everything I have to get done and how there's no way it's all going to happen and I've been up all night sanding the varnish on the wood menus with fine 220 sandpaper and then started hole-punching the laminated cocktail lists so I could lace them onto the wood, but I broke the hole puncher and thought I had a backup, but it was already the second one I'd blown through and it was after 10 p.m. so all the stores downtown were closed.

"So you need a hole puncher," Jeff interrupts me.

"Yes."

"I've got a hole puncher," he says as exhaustion slams into me and tears fill my eyes.

Jeff flies down the day before we open and crashes on my floor (Sasha's on the couch). He helps us punch the menus full of holes and lace them up and on opening night he runs the door.

It is the first of many reminders that success in this business takes a village.

---

* Author of one of my favorite cocktail books, *The Art of the Bar*.

CHAPTER FOUR

DILUTION AND CHILL*

My whole first year at The Varnish I'm alternately bone-tired and amped to the rafters, vampire-pale from living in the dark and skinnier than usual from all the cocaine helping to fuel my seven nights of service and seven days of maintenance. I have no space for anything or anyone outside the bar. The only time I see my friends is when they stop in for a drink. Courtney and I communicate through notes left on the kitchen counter and via text. Our infrequent face-to-face communiqués take place at The Varnish when she occasionally saves our asses by covering a shift. So as my parents and I sip our pre-Thanksgiving cocktails in the house they rented in the Hollywood Hills for the holidays, they ask if I'm enjoying owning a bar, I don't know what to say. It's easy to point to the challenges, but it's hard to put my finger on why they're worth it. I search for an example to give them, a picture to paint, but all I see are snippets of the past nine months on fast-forward. The

---

* This book was originally called *Dilution and Chill*—a title suggested by Jonathan Gold late one night in a bar in Santa Monica. Or Culver City. It may have been Venice.

overall is elusive. I can't catch it. I've never been good at examining something as a whole; I have to take things apart, piece by piece, in order not to get overwhelmed. When we were building the bar, every time I thought beyond the tasks I could accomplish on any given day I'd start to sweat.

My mother was the one who taught me to take things as they come. The first time I went to sleepaway camp I was scared shitless, terrified of being bullied. An entire month in a place living with people I didn't know? What would I do all day? Where would I sleep at night? What was I going to eat? I was freaking out, but my mother didn't shelter me; instead, she gave me tools to conquer. When I got to camp and opened my chest, I found a gift with a note attached. "Your clothes are organized week by week in layers. Each layer has a present. Don't rush through them. Enjoy each one as it comes."

So instead of trying to explain to my parents The Varnish as a whole—instead of trying to crack the meaning of what I'm doing, if there even is something so lofty as "meaning"—I think about yesterday and see if I can articulate.

Yesterday afternoon. I wake up late. A total anomaly. Courtney's side of the bed is empty; she's long gone, on set somewhere in Pasadena. I stretch out, luxuriating in my eight hours of solid REM before heeding the call of my bladder.

"How ya doin'?" I ask Freeway, sitting at my feet on the bathroom floor. Freeway is our cat who Courtney rescued after she was pitched out of a window on the 10 Freeway.

People are beyond fucking mental.

Freeway watches as I flip through my phone, checking the schedule to see who's on tonight.

"Matty's at the door, cool," I say out loud. Having a pet makes talking to myself seem much less weird. "He's great at keeping calm and organized which is essential because bars are fucking beasts around the holidays," I explain. "People traveling around in packs like wolves, howling instead of using their words and consuming more in one sitting than seems possible."

Freeway meows.

"Sorry, didn't mean to scare you," I apologize, flush, and let her hop onto the seat. She watches the water rush down and once it's settled, shimmies her butt off the edge, her claws digging into the faux wood seat I swapped in for the slippery white one so she could hold on. She tinkles successfully. I applaud her and we head to the kitchen for a treat.

A note on the counter from Courtney reads, "I made you a juice. It's in the fridge. I should be done with shooting by 7 p.m. and at the bar to cover that server shift by 8 p.m. Try to go to yoga like you said last night."

I make coffee, trying to remember when I said that. Courtney left for set at 5 a.m., so maybe when she got up I mumbled something of the sort. But she's right—yoga will be good for my brain and limber me up for tonight. Except doing yoga in L.A. is a three-hour event—driving there takes forty-five minutes. Class is an hour and a half, and if I can get out of there without too much fraternizing, I can be back home in an hour. I can't afford that kind of time suck today since I have to be in the bar at 4, so I throw down my mat, pop in a Rodney Yee DVD, and do fifty minutes in the loft with Freeway watching closely.

BY 3:45 I'M OUT OF the shower assembling my battle gear. In the theater, I was taught to rehearse in the shoes my character would be wearing in the play. With the ground as a constant point of contact, every step informed my posture and the way I moved—heavy or light, fast or slow—revealing my character's behavior. I'll be in character tonight for seven hours with the aim of projecting grace and a generosity of spirit, a little peacocking, a lot of confidence, and I need to move with ease 'cause that's a long time to be onstage.

I choose one of the bespoke suits I bought from the tailor we used at Little Branch—Patrick's Fashionway in Bangkok. Patrick himself showed up at The Varnish six months ago to take my measurements, and eight weeks later, two three-piece suits and two shirts arrived rolled up in a DHL package. These beauties take a sound beating behind the stick—stains from bitters and cocktail splatters that will never come out—and even though they're already threadbare, they look great in low light.

Gray slacks and vest. Blue button-up shirt. Cuff links. Pink paisley tie. Garters that attach to the tail of my shirt and the tops of my socks so my shirt doesn't slip out of my pants when I'm shaking, and arm garters to keep my shirtsleeves tacked neatly above my biceps. On my feet, a pair of Vans, partly because they're comfortable and partly because they make me feel a little punk rock.

I do the phone/wallet/keys check at the door—all there—and then hop on the service elevator to the basement garage, jam my keys into the bar's back door, and head into the office, where I drop my shit on the desk—three planks of wood pushed up against the wall—and pull the frozen pans of ice out of the chest freezers. This is our daily harvest and one of the most important jobs for setup. There's never a time the early bartender doesn't pull, harvest, and refill every one of the forty ice pans. Harvesting ice is our version of a chef pulling their produce out of the ground, getting intimate with the shapes and densities of the ingredients they'll be working with.

While I wait for the pans of uncut ice to come to temp so I can chop them into rocks and spears, I carry the already cut pieces left over from last night to the bar and stock the chest freezer, then drop spears into the racks of collins glasses in the glass frosters—a tiny step that streamlines service. Especially on busy weekend nights, it's a real time-saver.

Carlos arrives on time, as always.

"Hola Papa," I say. "Está bien?"

"Si Papa."

"Sandra bueno?" I ask after his wife in crappy Spanish.

"Muy bien. Más tacos," he strokes his belly, then immediately starts squeezing juice with the Ra Chand manual juicer that towers over his head. I hoof it back to the office, where the ice is ready for carving.

Popping a block out of its plastic hotel pan, I place it in a bus tub that has a plastic cutting board lining the bottom. I hold the block steady with my chisel, finding the invisible line I want to follow, and tap the top lightly with a small hammer. The play is to hit it with the perfect amount of force so the cut is clean and the chisel slides through the ice like butter. Ideally I'd extract six spears from the slab, but that only happens when I'm really in the zone. Usually I get four or five, and the odds and ends get used for chilling stirred-up cocktails. The rocks are harder to be exact with since the ice slabs are thicker—the pans for spears are filled with 750 milliliters of reverse osmosis water, while the ones for rocks get 1,250 milliliters. Today I get a stock of eighty-five rocks, with plenty of odds and ends for cracking into stirred cocktails, and ninety-eight spears—an amount I'm happy with.

Now I need to eat, and since I'm feeling health-conscious thanks to a full night's sleep and yoga, I slip out of the bar and head two blocks down to Blossom, a family-run Vietnamese restaurant where a young woman who is probably the owner's daughter, because she's *always* there, takes my order. To me, this place makes the cleanest, lightest, most satisfying food. Every item on the menu is fresh and delicious and simple. I love simplicity. It's what I strive for at The Varnish, and in life.

→═◦ ◦═←

BACK AT THE VARNISH, I'M shoveling my order of bún* down my throat out of one of the two stainless-steel bowls I brought in for this purpose—no matter what point you're at in your hospitality career, if you eat at work, you're always doing it in a hurry—when Kimmy enters through the back curtain with her standard "I'm here bitches!"

Matty follows right behind looking like Kramer, fumbling with his tie.

"Can I?" I ask. Then whip it into a double Windsor sans mirror—a trick I learned from watching my dad every morning for eighteen years.

"We all good in front?" Matty asks.

"Just set us up with a family meal and we're off," I say, folding his collar down.

"Doctor's orders?"

"Mezcal for me."

"Rye for sweetie pie!" Kimmy calls from behind the door, where she's putting the finishing touches on her outfit—a nifty mix of Lucille Ball and the Roaring Twenties.

"Naranja, Papa," Carlos adds.

"Where's Bostick?" Matty looks around.

I shrug just as two meaty hands part the curtain. "I have arrived," Bostick proclaims. "The brunch was legendary."

Bostick speaks in Bostickisms, most of which do not require a response.

"You ready for tonight?" I ask.

"Just another ride on the roller coaster with my alligator arms up."

---

* Vietnamese grilled-meat-and-vermicelli-noodle salad.

I assume this is a good thing as we file out of the office and wend our way through the bar. I pull the dimmer switch down to hit the Sharpie marking. Kimmy swipes at a crumb on the table with a rag. Matty whisks a menu off the bar rail and Bostick flicks on our up-tempo Northern Soul playlist to start—it's the night before Thanksgiving, which means everyone's out getting lit, so there's no need to "ease" in.

"Ojos y'all," I say as we toast our family meals and eye-fuck* each other. "And remember—we are craftsmen," I pontificate because I'm feeling emotional. Tomorrow's the first night we've gone dark since opening. "We aren't artists, mixologists, or bar chefs, just bartenders, improving our craft."

Sometimes the things I say sound like they're ripped from the pages of a self-help brochure, but in the inimitable words of the Beastie Boys: *Be true to yourself and you will never fall.*

"Super deep," Kimmy grins.

"Now can I open the door?" Matty asks.

"Flip the lock," I tell him, and an entire evening awaits. It might be an evening of joy! It might be an evening of sorrow. It might be an evening friends come in and light the place up and it might be an evening the HVAC breaks because I forgot to "accept" the maintenance invite. But no matter what kind of evening it is, one must always consider it's going to be someone's first time, and they deserve the best.

This aphorism I keep to myself.

BY 9:30 THE TABLES ARE packed, we're two deep at the bar, and I'm wondering where the fuck Courtney is when she sends a text

---

* When you look pointedly into everyone's eyes after a toast and before drinking.

saying she's on double overtime on set which is major moola so she can't leave. Bostick is in the service well, and I'm in personality, where I'm hoping Carlos will soon materialize with a pan of shaking rocks so I can finish this round for the couple standing in front of me. I communicate via facial expression with Matty at the door to pump the brakes. It's a combination wide eyes + shake of the head. We're at a perfect capacity and Bostick and I are in sync back here, each of us moving around the other in a choreographed flow, effortlessly carrying on conversations with our guests, joking with each other, floating bottles back and forth, shaking tins, cracking ice, spinning labels customer side out, and enchanting everyone by pouring the liquor so skillfully it hits the line on our jiggers every time.

Bostick dry-shakes a Ramos gin fizz with short, hard strokes while I stir up a manhattan in a mixing glass, like spinning a spool of liquid silk, the rye's dark legs spiraling down the glass, lightening in color as it dilutes. We swap movements at the same time without a word and Bostick stirs two drinks with one hand while prepping a third with his other, while I start shaking with ice—a sound aggressive and loud at first that winds down to a dulled hum as the edges soften and become round. Every action, every move, every step has a purpose, but yeah, we're onstage, so we imbue our purpose with a little flair.

"Eek! Where's Courtney?" Kimmy squeaks as she sails by, whisking up the tray I just finished expediting. Thankfully she doesn't wait around for an answer—I don't have the heart to tell her it doesn't look like Courtney's going to make it and she's going to have to manage the floor on her own.

With twenty-four drinks on the board and the personality well covered, I jump out from behind the bar and grab the full tray on Bostick's side. Devon Tarby, a cocktail server and daytime bartender at Asia de Cuba in West Hollywood who's been coming here since the day we opened and quietly observing, clears out while I squeeze by.

"Hey D., can I get you a drink after I run this?" I ask her.

"I'll just do some white wine. Keep it easy for you," she says.

"Ha. Appreciate that. Hey, is it warm in here?" I dab my forehead with my sleeve. "I can't tell; I've been moving nonstop."

"It's not too bad, but it's packed, so maybe a little air. The lights are good, but I think you should turn up the music a touch."

Impressive how attuned she is, I think, as I hop over to table 9 with my tray of drinks glimmering by candlelight. Upon arrival I notice there are no napkins.

"Here you go boss," Devon appears like a rabbit out of a hat with a stack.

"*Fuck yeah*," I mouth, and place the napkins down. Then, starting at the bottom left of the tray and working my way clockwise around the table, I set down a negroni, a whiskey sour, an old-fashioned, a lager, a Gold Rush, and a Floradora. Get some oohs and ahs with a few camera phone pictures and an accidental flash.

"Enjoy them while they're cold and frosty!" I say, hoping to quash the photography.

I jump back into my well, grab one of the tickets piled up in front of Bostick and dig in; a metronome in my head keeps me on point, silently singing the names of the drinks I'm making in my head. Every chef, bartender, and barista has their own technique for building rounds, but it's definitely instinct over intellect.

Plane, Mule, Lady, GFC. Plane, Mule, Lady, GFC. Plane, Mule, Lady, GFC. Order up!

I can hear snippets of conversation from every corner of the room. *Hi there! Oh, it's your first time? It's my birthday! The Lakers won? Are you waiting on a table? I live upstairs . . .*

My tray of drinks is dying and I can't get Kimmy's attention and I guess I'm looking distraught because—

"I can do it," Devon says. "I can run that tray."

"Uh, are you sure? Do you even—"

"Let me give it a shot." She grabs the tray from the sides and the drinks wobble.

"One hand on the bottom," I tell her.

She moves her hand underneath and everything steadies.

"It's table 7, six drinks, clockwise from here on the tray and then served clockwise at the table starting with the first person on your left."

"Got it," she nods, and lifts the tray off the pizza stand,* setting a replacement one down with a votive in the center, setting up whoever runs the next round.

"About time that happened," Matty calls out, watching Devon make her way calmly through the crowded room. "She's been watching us for months."

AFTER A FEW MORE ROUNDS there's a natural pause. This is when you can easily fuck yourself and cut someone loose, thinking the night's winding down, but it's just a beat before getting your second slaughter.

"Damn, if you let go of the reins she will buck you off," Bostick grins.

Like most things he says, it takes me a minute to figure out.

"I saw you sweating a few times through those stacks," I tell him.

"Nothing but throwing a gallon of water into a lake."

I really don't know if Bostick's a genius or insane.

---

* We repurposed pizza stands to hold two cocktail trays—the elevated tray at the top makes it easy for the server to slip their hand underneath and puts the cocktails on view. The empty tray on the bottom is moved to the top before the server leaves the service station, setting the bartender up for his or her next round.

"I'm gonna take five," I tell him. "You got this?"

"I'm like a freight train—hard to get going and hard as shit to stop."

"Right," I say. "I'll take that as a yes."

In the office I do a couple leg stretches, then take my life into my hands and sit down on the stool for a breather. The door to the office swings directly into the stool and opens constantly, since the office doubles as our back of house, so sitting down is never a good idea. But it's the only chance I'll have all night. I think maybe I'll look at the news, see what's happening in the world, and pull the top of the laptop up to find a baggie of blow. I pause. I could take a bump, but the sleeping in, the yoga, the healthy dinner . . . I uncharacteristically place the drugs back in their "hiding spot" and walk out of the office. Peeking through the curtains before heading back into the fray I see Kimmy, who despite being the only server on the floor has things locked down. There's a celebrity sitting on the piano bench next to our pianist Jamie Elman, singing. Everyone in the bar is trying to pretend they don't notice. Carlos is refilling water glasses. Bostick is welcoming new guests with an engaging smile, and Devon is helping Matty seat a party of six.

I'm going to hire her tomorrow.

BACK IN THE HOLLYWOOD HILLS, I gaze out through the floor-to-ceiling windows at all the houses wrapped in holiday lights and tell my parents, "What I love about The Varnish is that it's a family." And then I turn to them and smile so they understand they could never be replaced. "Ma deuxième famille."

CHAPTER FIVE

COLD AS ICE

Ice is a big deal.

Like a sandwich shop baking their own bread or a coffee shop roasting their own beans, we freeze and cut our own block ice for The Varnish. *But why?* people ask. *Isn't ice just frozen water? How come you don't use regular machine-made cubes?*

Well if approximately 25 percent of a drink is composed of melted ice, aka water, wouldn't you use the best you could?

Consider ice the bartender's flame—it's what "cooks" a cocktail— and the various shapes and sizes are its settings (high, low, or simmering), each one allowing the drink to achieve its ideal dilution.* The moment ice is introduced, the life of a cocktail becomes a battle of durability and longevity, with drinks served *up*† being

---

* Many say that a well-made cocktail should contain 75 percent ingredients and 25 percent water melted from ice. This is a generalization. A bartender should regard the style of cocktail, as well as the ABV (alcohol by volume) of the spirit they're using, to accurately address water content.

† A cocktail shaken with block ice and strained into a chilled coupe, or a cocktail stirred with cracked block ice and strained into a chilled coupe.

more fragile than drinks served *long** or *down*.† The idea is to get the drink, in as cold a state as possible, to the guest. If you have only one drink in an order, you can just run it out; not so difficult. But things become trickier when you're building a round and have a number of different styles of drinks to contend with. It's like food— the moment something is out of the oven, it begins to lose heat; in some cases it may start to wilt or deflate. If you're serving dinner and want every item to reach the table hot, it's imperative to know which you can pull out of the oven first—which are more durable— and which you want to pull out last because they're more fragile. Same thing goes for cocktails. From most durable to least, here's how cocktail styles stack up:

→ SPIRITS SERVED NEAT: bulletproof, since there's no ice to melt and become added water.

→ OLD-FASHIONED/ON THE ROCK-STYLE DRINKS (drinks served over one large rock): durable, since there are only large surfaces of a single melting cube to affect them.

→ TOM COLLINS-STYLE AND SHAKEN DOWN DRINKS (drinks shaken with ice and served long over a spear of ice or down on a rock of ice): less durable.

→ CRACKED-ICE COCKTAILS (drinks served over hand-cracked ice): even less durable because of the additional surface area of the many pieces of cracked ice.

---

* A shaken cocktail with citrus juice, sugar, and a base spirit, strained into a tall twelve-ounce glass with a single long spear of block ice and finished off with soda water. A long drink can also simply be a base spirit served with bubbles and built directly in a tall glass. Long drinks are usually referred to as highballs.

† A cocktail shaken with block ice and strained into a chilled double rocks glass with a single chunk of block ice. Also an old-fashioned cocktail built in the glass over a single rock of ice.

→ FIXES/JULEPS/SWIZZLES/COBBLERS (drinks served over pebbles): fragile, due to all the surface area of the crushed ice.

→ "TRADITIONAL" SOURS (drinks shaken with egg white): more fragile, having only minimal ice crystals from being shaken and served without ice. Note: the sour is slightly more durable than the daiquiri (see below) in texture and appearance, due to the binding properties of the egg white.

→ DAIQUIRIS, GIMLETS, AND "NONTRADITIONAL" SOURS (drinks shaken with ice and served straight up): even more fragile.

→ MARTINIS AND MANHATTANS (drinks stirred with ice and served straight up): the most fragile, since they have no ice crystals, citrus pulp, or air bubbles to help retain their chill, nor are they served over block ice, so they will grow warmer the fastest.

As bartenders, our goal is to have as much control over our flame as possible in order to make the best drinks possible. The size and durability of block ice helps us execute cocktails with precision to achieve the correct temperature and water content, and also to make them look fucking cool.

At The Varnish, we use block ice because our cocktail program is based on techniques tracing back to the 1800s. At that time, block ice was harvested from shallow, slow-moving lakes, ponds, streams, and sometimes rivers by teams of men and horses. Locations were chosen where the ice was strong enough to hold their collective weight so no people or animals would go crashing through as they worked, cutting out blocks of ice using a combination of horse-drawn "ice plows" and crosscut hand saws. They would then wrap the ice in hay to impede melting and float it through a channel they'd cleared toward an icehouse for storage or onto a train for shipping. Richard Boccato—owner of Dutch Kills in Long Island City, a sister bar to The Varnish—

illuminates further in his meticulous treatise on ice, "The Cold War."[*]

> Ice crystals form in any body of water at what is called a nucleation point—a piece of foreign matter around which an ice crystal builds, usually the product of an impurity floating in the water. In the example of a lake or a pond (each of which tends to freeze from the outside in), nucleation points happen near the shore where the water is not as deep, and the initial ice crystals aggregate in nonlinear formations across the surface. As the ice crystals form, the lattice becomes so tight that there isn't room for impurities to make their way inside, which forces them to be pushed away.
>
> Unlike slow-frozen lake ice, most home-frozen ice cubes develop a cloudy center, because the water is freezing on all four sides from the outside in and from the top down, trapping air bubbles and impurities in the middle. When the last bit of water on the inside of the cube turns to ice, small fissures and striations develop and branch outward from the core, creating countless miniature fault lines. The impurities and oxygen bubbles then attempt to forcibly push their way out from the core as the water expands during the freezing process, lending the cube an overall "fuzzy" aspect.
>
> Because water in a lake or a pond freezes from the top down, oxygen bubbles and impurities remain unfrozen together at the bottom. As the ice thickens, the water below becomes ice at a slower rate. (Ice is a poor conductor of heat.) The thicker the ice becomes, the further the latent heat (the amount of energy required for water to change state) that is released during the process of freezing has to travel to reach the cold air above the pond. In essence, the thicker the ice, the slower the rate of freezing—and melting.

---

[*] Published in Gaz Regan's *Annual Manual for Bartenders,* 2013.

A fun way to test this theory at home is by taking a handful of machine-made ice and a two-by-two-inch rock of block ice and storing them in the same freezer so they come to the same temperature. After thirty minutes, place each ice style in a glass with two ounces of eighty-proof booze. Let them sit for fifteen minutes, then take a look. I guarantee the machine-made ice will have broken down far more than the single large rock, introducing more water—i.e., dilution—into the drink. This is because machine-made ice has proportionally more surfaces for heat transfer to occur, and also it's produced in molds injected with cool air, which helps speed up the freezing process but renders the ice less dense.

When it comes to using ice for mixing drinks, it's important that it be a combination of dry and clear—any surface melt or sweat[*] introduced into a shaking tin or mixing glass will add water and result in an overdiluted cocktail that will never hit the ideal temperature.[†] And even though the droplets of water on the surface of a cube are 0 degrees Celsius, they have already lost their chilling power. The thermodynamics of chilling a boozy concoction happens by heat of fusion, which results from providing energy, typically heat, to a substance to change its state from a solid to a liquid. This change is measured in "science calories." Warming up a gram of ice from –1 Celsius to 0 Celsius takes a small half calorie (0.5) of heat energy, but going from 0 Celsius to melted takes *eighty calories of heat energy.* And by "heat" I don't mean a hot flame, but the heat energy from the movement of molecules. It's at this state change that the chilling power of ice is released into the surrounding liquid.

---

* The moment ice is exposed to the elements, it starts to absorb heat and gently melt.

† At The Varnish, the desired temperature of a stirred cocktail is –8 Celsius (17.6 Fahrenheit), and a shaken cocktail is –6 Celsius (21.2 Fahrenheit).

Another crucial factor is the starting *temperature* of your ice. In his seminal book on the element, *Liquid Intelligence*, Dave Arnold says, "All ice is 0°C or colder, but it can be much colder or barely colder, and if you're making cocktails, barely colder is better. Cold ice shatters and is no fun to work with, while ice that has warmed up to freezing temperature is a pleasure."

This is why we store aluminum pans of respective ice styles in Perlick glass chillers set to –10 Fahrenheit behind the bar, as opposed to filling wells[*] with blocks that are open to the ambient temperature of the room. When building rounds, we remove a pan of ice from the Perlick and place it in the crushed-ice bin to allow it to temper, so it doesn't experience thermal shock when added to liquid. We then consistently monitor this ice to ensure it stays in the range of "dry and frosted" moving toward "clear and wet," but definitely not dripping. Any pan full of ice that becomes wet and dripping gets swapped out for a fresh one.

When we stir a cocktail, we crack[†] the block ice into smaller pieces and place them in a chilled mixing glass with the ingredients. The cracked pieces have much less surface melt than machine ice sitting in a well, since we crack them off a block immediately before making the drink. This allows us to create super-cold cocktails with a silky rather than watery texture.

---

[*] Deep sinks filled with ice and lined with a speed rack to hold liquor bottles. The name is derived from pre-plumbing days, when water had to be fetched by hand from nearby wells, lakes, and rivers. These sinks were originally called farmhouse sinks or apron-front sinks since they were designed for women who spent long hours toiling away cooking and washing. The ergonomics of the new sinks eliminated the countertop that previously made users lean forward and ruin their aprons.

[†] Using the back of a spoon or the blunt end of an ice pick, we "crack" ice off a large block.

We shake cocktails with a tempered piece of block ice, which adds the necessary aeration and dilution to arrive at a frothy texture. That single smooth rock has far fewer nucleation points, aka those imperfections that promote aeration. If you use machine-made ice, with all its edges and corners and dings and cracks, you'll get a bunch of nucleation points, which when you shake, will break down the ice and degrade the ingredients faster.

The only machine ice we use at The Varnish are called pebbles,* which are added to drinks demanding crushed ice, like fixes, juleps, swizzles, and cobblers.

On top of all the science-y reasons to use block ice, there's the sound, which was my first introduction. I didn't see it; I heard it. The *cla-clunk, cla-clunk, cla-clunk* of a rock in a shaking tin when I walked into Milk & Honey. The *thunk-ssshhhh* of someone cracking the back of a spoon against a block. The *snap, crackle, pop!* of ice shards floating on top of my daiquiri . . . It's a beautiful symphony when you're a customer or behind the bar, a percussive conversation. I can tell where my bartending partner is in his or her round just by listening. There's also the indisputable beauty of crystal-clear ice, holding a negroni up to the light and being able to see clear through it to the other side of the room.

I mentioned before that we use block ice because our cocktail program is based on techniques like ice-harvesting in nature that predate ice machines and freezers.

One of the earliest known examples of ice harvesting is from 1806, when a twenty-two-year-old man named Frederic Tudor, known as Boston's Ice King, harvested ice from Thoreau's

---

* Ice that's slightly more durable than straight crushed ice and very pleasing to look at. Ours is made in the Scotsman NME654.

Walden Pond and shipped it off to the Caribbean, Louisiana, and India to "cool drinks, preserve food, and soothe patients suffering from yellow fever." Thoreau responded to this pilfering in *Walden,* writing that "the sweltering inhabitants of Charleston and New Orleans, of Madras and Bombay and Calcutta, drink at my well."

The next most significant historical ice occurrence happened in 1851, when John Gorrie, of New Orleans, submitted the first model of a mechanical ice-making machine to the U.S. Patent Office. It was designed to "convert water into ice artificially by absorbing its heat of liquefaction with expanding air." The public found his invention preposterous, and Gorrie was made a mockery of, dying four years later a self-professed failure. If only he could have sucked it up another decade or so, until 1868, when the first commercial man-made ice plant opened in New Orleans using similar technology, he could've given all those haters a big "fuck you."

By the 1910s, ice plants were popping up all over the place, and the days of lake harvesting had disappeared. Manufacturing commercial ice became a booming business, with hundreds of storage facilities constructed along the railway lines to give easy access to train cars. These plants froze ice in massive basins, producing rectangles larger than a basketball court that were then scored and crushed and shipped off to their buyers.

Which brings us up to the counterculture brouhaha of the 1960s, when a man named Virgil Clinebell, living in Loveland, Colorado, got the idea to purchase the blocks from the plant, then crush or cube them himself and sell them locally. Clinebell first wrapped his deliveries in wax paper before graduating to plastic bags originally used for carrots, that were washed out, filled with ice, and tied off with blue twist ties. "We spent our childhood summers twisting those ties until our fingers turned blue," Mike and Brenda, Virgil's children told me.

But in the early 1980s, the EPA started shutting the large operations down when it found that the ammonia used in the lines to cool the large vats was leaking into the ice—an assured health hazard. The loss of his supplier compelled Clinebell to develop the CB300X2D*—a block-making machine that is fast becoming an industry must-have for those serious about their ice.

WHEN THE VARNISH FIRST OPENED, we made our own ice like they did at my friend Richie's spot, Dutch Kills—by slow-freezing reverse osmosis water in four-by-six-by-twelve-inch clear polycarbonate hotel pans and storing them in chest freezers overnight. The following day, the lion's share of the bartender's prep was harvesting the ice with hammers and chisels before service—a two-hour endeavor at best. The results were pure, but not completely clear due to the frozen oxygen bubbles and minerals within the ice, which lent a fuzzy aspect to it and caused fractures in the crystalline structure, making it less durable. At some point, Richie made the shift from freezing water in hotel pans to ordering crystal-clear blocks from Okamoto—an ice-sculpting studio that used Clinebell machines. Eventually, Richie bought his own Clinebell and started producing crystal-clear product on-site. Once friends in the industry caught wind of it, they asked him to make ice for them, and soon enough, Hundredweight Ice was born.

---

* The CB300X2D is the largest model, with two basins measuring 48.5 inches wide, 51.25 inches deep, and 42 inches high. When each basin is filled with 40 gallons of water, it produces two crystal-clear 300-pound blocks that are 20 × 40 × 10 inches thick.

It is said that imitation is the sincerest form of flattery, and in 2012, I followed in Richie's footsteps, building my own ice company in L.A. called Penny Pound Ice with my partners Cedd Moses and Gordon Bellaver and a dedicated team headed up by Hector Lopez-Flores. While we didn't have to harvest ice from lakes, fingers raw from the cold and brains anxious over the collective human and equine weight, the process of cutting and delivering ice was as analog as it would ever be. It was just me and Hector in the guts of a trilevel warehouse building with two Clinebells, one chainsaw, one bandsaw, a chest freezer, a drain, and a wet/dry vacuum. To reach the space, we had to walk through a large parking lot in the back of the warehouse and ride a rickety service elevator one flight down, where it opened onto a large storage space lit by a bay of buzzing fluorescent lights. Old TVs and amplifiers lay scattered around the floor in various states of corrosion. We created a "room" in the middle of the mess by hammering together a few walls out of lumber and attaching a chain-link fence. Our own frozen version of *Breaking Bad*.

We started out making ice for The Varnish and Tony's Saloon, and within six years were serving 175 clients. By then we had built a proper factory, which is what we operate out of today. With a total of twenty-six Clinebell machines, we can spit out eighty to a hundred three-hundred-pound blocks of ice every week. Each block is hoisted out of a Clinebell and ripped down by a chainsaw into eight "loaves" of ice, which are then passed on to the bandsaw stations for the smaller hand cuts—around 400 to 750 individual pieces, depending on style and size. These cuts are organized onto baking sheet pans and slipped onto a rack in the walk-in to refreeze. Once they're ready, each piece is packaged into eight hundred bags, fifty units per bag, by Aldo, our youngest team member, who's still in high school.

On average, we hand-cut about forty thousand individual units per week.

FOR TOO LONG, BLOCK ICE has been considered a fancy ingredient found only in craft cocktail bars. The fact that it's enjoying a place in more and more spots of every kind is a testament to not only how cool it looks, but how essential it is to making great cocktails. Having Penny Pound has allowed me to be part of this critical renaissance, to create something I believe adds substance and style to the world of hospitality and injects a little bit of analog into our overly digitized lives.

THE GREATEST HITS AS PENNY POUND ICE SEES IT

 THE ROCK: The Rock is our bread and butter and was designed to be larger than a standard two-by-two-inch because it needed to sit taller in the glass but not hit you in the face when you took a swig. Ideal for all drinks on the rock or shaken down.

 THE STONE: Perfect for a sipping spirit or smaller glassware.

 THE SPEAR: Best for drinks served long and tall.

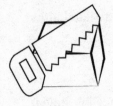

THE CINDERBLOCK: For the DIY bar, bartenders can chip pieces off the block, giving their ice an irregular, handcrafted appeal.

THE NORMANDIE: Bigger than the Rock for those larger-than-life glasses. Named after the Normandie Club,* which had bigger glasses and needed a fatter piece of ice.

PEBBLES: Use in juleps, mojitos, mai tais, cobblers, and fixes.

---

* Opened in 2015 by Proprietors LLC and 213 Hospitality, a neighborhood cocktail bar in the heart of Koreatown, housed inside the landmark Hotel Normandie.

CHAPTER SIX

HOME DEPOT . . . ONE YEAR LATER

A barista hoses down the sidewalk outside Spring for Coffee on Spring Street and Sixth, pushing the effluvium of last night's doorway dwellers into the street. I step carefully over the watery muck and duck inside. Order an Americano and lean up in the corner by the edge of the milk-and-sugar rail to await my morning hair bender. My rule about not reviewing emails before caffeine goes to the dogs since I'm running behind, so I take a peek at last night's close-out report: one broken glass washer, one cracked and now sporadically functioning iPad, three menus turned to kindling thanks to a pyro guest, and the loss of our cocktail waitress Georgia, who has a chronic problem getting to work on time.

The "Go Ahead and Vent" section further illuminates.

→ "I had the iPad on my lap for a second and reached for something and it slipped and fell and cracked a little bit in the corner. *Sorryyyyyyy*."—Sari

→ "Just because we're busier doesn't mean we're making more tips."—Gordon

→ "You would think people grow up after high school. Thursday nights continue to challenge that concept."—Ky

"Eric Americano!" the barista singsongs.

I grab my coffee as the rain-soaked strains of "Riders on the Storm" kick in and walk out onto the street to find Sixth and Spring locked down by a film crew. A PA* stops me, and in the most Cali of lilts says, "Hey bro, can you wait until they call 'cut' and turn off the rain machine?"

Fucking Hollywood, I think as the fake precipitation falls, and I remember a time not long ago when pools of girls, bowls of drugs, and precipitously balanced Case Study Houses filled with positive vibes meant I was living the dream. But more and more lately it feels like a nightmare. As a bartender, my responsibilities used to end when I locked the door behind me—I was free to stay up all night! Sleep all day! But there is no such luxury for a bar owner. The responsibilities never end, the door never fully closes; it's always cracked, the dim glow of the drinks cooler spilling into the furthest recesses of my mind. It's the same for my friends with restaurants—there is no such thing as "downtime" because something is always happening or about to happen or finishing up just having happened. It's likely the same for any business owner, except in hospitality nothing is ever "over." In most industries, the projects to be done have beginnings, middles, and ends. Sometimes those beginnings may get pushed back and the ends extended, but in hospitality the project, once up and running, runs forever. It's like doing a play that never stops, but in the theater at least Mondays are dark. There are no dark nights at a bar, which is amazing

---

* Short for "production assistant"—the lowliest position on a film crew.

because it means I get to go onstage *every night* and practice my craft. I just didn't stop to think about the daytime responsibilities. I knew there would be an inherent deck of uncontrollable variables I'd have to deal with when we first opened—from budget concerns, to Cole's GM considering my classic cocktail bar in the back an affront to his established rock-and-roll bar up front and calling us uptight assholes in suspenders, to hours spent peeling protective laser film off refrigeration equipment before plugging in the units but only after waiting twenty-four hours to let the refrigerant oils settle so I didn't brown out the compressor*—but the steady flow of daytime responsibilities never crossed my mind. Even though I worked in the industry, I was blinded by the lights, imagining that owning a bar would be some Rat Pack–level shit—me in a dimly lit room in a beautiful suit, greeting guests, kissing cheeks, shaking hands. I never once considered that all the minutiae allowing me to stand onstage and perform nightly would fall to me to attend to.

As it turns out, just like every other glamourous job, owning a bar has a very glamourless side.

"*Cut!*" yells the director, swaying in the bucket seat above my head.

---

* Side effects of opening a bar may also include cleaning up vomit, stitching ripped leather booths, putting out napkin fires, snaking bar sinks and clogged floor drains every two weeks, ordering replacement barware as it mysteriously disappears, finding one of your custom-made wooden bar menus on the bookshelf at a "friend's" Christmas party, 5 a.m. security alarm phone calls after a Skid Row tweaker puts his fist through the front door, rushing your server to the ER for a peeler incident, discovering the burning-eye sensation your staff and guests are experiencing is a result of the HVAC system you share with Cole's distributing lachrymatory-factor synthase from chopped onions into the air, an ex-girlfriend sharing a drink with your present girlfriend while you're bartending, leading to an ill-considered threesome, general fatigue, nausea, occasional rectal bleeding, heart palpitations, and perpetual insomnia.

The fake rain stops.

The actors take five.

"You can walk now," the PA tells me.

"Thanks man," I say, and pull out a business card. "If you guys are looking for drinks when you finish, we're right up the street." I point toward Main. Add a drink token because being a PA sucks.

"It's your place?" he asks, eyes wide.

"Yeah," I sigh. "It's my place."

OUTSIDE OF THE VARNISH, Home Depot is where I spend most of my time. While I can't weld pipes or build fences, and the one time I worked in an actual carpenter shop I was sent home after two hours because I couldn't figure out how to make a proper wooden rectangle, I love the loamy smell of Home Depot's garden center, finding just the right screw out of the thousands of drawers filled with thousands of screws, and how, if you stare long enough at the grain on wood you can see distinct shapes and faces.

Potentially the result of all the acid I did in high school.

That said, I totally get how Home Depot is a certain kind of hell for people: the too bright lights; the combined aroma of hot glue, paint thinner, and metal; the way it resembles George Romero's *Dawn of the Dead*—the one where all the people take refuge in the mall from flesh-eating zombies. But to me, the fifty-four aisles of Home Depot #1048 in DTLA is pure heaven, and the minute I step inside I go full *Rain Man*.

In aisle 1, I grab stainless-steel cleaner for the Perlicks which get messed up from the mineral deposits in condensation and the citrus, sugars, and booze constantly coming into contact with them.

My phone buzzes.

Papa! Can you text me the guy who fix walk in again? Freezer not working.—Hector (Penny Pound Ice manager)

I send Hector the number and slide over to aisle 5 to look for a switch guard to screw over the light panel so the light switch stops getting knocked off when someone walks by on their way to the office. It doesn't happen often, but when it does, all the lights in the bar go off. Part of me thinks I'll miss those moments—the whole room raising their voices in concert, the bartenders swearing, the servers yelling . . . I'll also miss it as a signal that someone is headed to the office, which happens for only one of three reasons: to grab cigarettes, do a bump, or engage in some form of sexual gratification.

I toss the switch guard in my cart—a two-pack for $1.99—and head further down the aisle toward the lightbulbs. I am forever buying lightbulbs. They blow out from power surges and from staying on nearly constantly. Even if staff remembers to turn every light off at the end of service, the cleaning crew throws them back up at 8 a.m. I grab backups because any day now California is going to outlaw filament bulbs and require commercial establishments to use LEDs, and while I'm all for saving the environment, the temperature on non-incandescent bulbs is stale and static and I hate it.

My phone buzzes.

The comp tab you requested for that chef hit $250 . . . maybe next time we just send shots?—Max (The Varnish GM)

The hospitality industry has a supercool brotherhood that silently acknowledges that if you work in it, you drink and eat at a discount or get stuff sent out from the kitchen or bar for free. It mostly happens in neighborhoods where everyone knows everyone else and we all hang out in each other's spots. But I've encountered the camaraderie far beyond my neighborhood and will tell you that there are few more life-affirming experiences than being in a foreign country and getting gifts sent over by the chef or bartender.

Usually at home, a round of snaiquiris* or a plate of snacks or a few "bartender handshakes"† is sufficient to show your love, but there are nights when a chef comes through who destroyed you when you ate at their spot and you want to pay them back, so you give them an open tab. It's often a game of chicken to see who can be the host with the most, which can turn into a never-ending pounding-on-the-chest battle of the most excessive kind.

My phone buzzes.

Oh, and can you swing by the bank and get singles and fives for tonight? I'm stuck waiting on Champagne delivery.—Max again

I add "bank" to my to-do list and glide over to aisle 6, which thrills me with its multicolored zip ties which I will use to gather all the cables hanging out of the backs of various devices—and getting tangled unattractively—into one neat bundle. In aisle 12 I find the plumber's tape, which I need to wrap around the threads of the pipe when I screw on the replacement faucet in the women's restroom, which was ripped off last week. It's one of those things that happens semi-regularly and never ceases to amaze me. You have to be *incredibly inebriated*—I mean past the point of no return—to rip that thing off. I sometimes wish alcohol was more like mushrooms, and that instead of jacking people up, it chilled them the fuck out. Made them want to touch and caress things instead of tearing them to pieces.

My phone buzzes.

New review on Yelp! Someone on Yelp has just reviewed your

---

* A half daiquiri.

† Shots to say hello, shots to welcome someone else, shots before leaving.

business. Click here to access your business account and read the review.—Yelp (for business owners)*

Now I know I shouldn't look at Yelp unless I'm sitting down with a cold beer in hand and soothing music playing, but I can't help myself. I stand in the middle of the plumbing aisle and hit the devil-red-and-black icon to read about a reviewer who googled The Varnish and found photos of sushi and karaoke (neither of which we offer), then visited for the *express purpose* of enjoying our classic cocktails accompanied by raw fish and Raw Power, only to be *wildly disappointed* that only one out of three items "advertised" was on tap. In response, they've given The Varnish one star because, okay, fine, the drinks were good, but everything else SUCKED.

"Fuck!" I say, and I guess I say it out loud because when I look up from my little black mirror, a mom is glaring at me and her child is laughing.

"Sorry," I mumble. Look back at my phone. Press RESPOND. Then stop and put my phone back in my pocket. Remember I need to frame a gracious response that explains to this person that we can't control what users post, and do it in a manner that will not only appease but convince them to take down their bad rating which has the power to directly impact our bottom line† and which, according to studies, if left unanswered, will signal to other customers we don't give a fuck.

Speaking of not giving a fuck, I need a replacement set of bar

---

* I hate the Yelp platform, mostly because reviewers use it as "Why me?" therapy. Sorry you didn't like the music or that the girl you were hitting on told you to suck it, but that has nothing to do with whether a bar is "good" or "bad" and is not helpful information when letting other people know what they might expect should they visit.

† Studies show that somewhere between 67 and 90 percent of consumers look at reviews before making decisions about what to buy or where to go.

keys because I lost mine the other night somewhere between the bathroom stall and the mystery bed I ended up in, so it's off to aisle 15. As the aproned dude behind the counter grinds my keys into submission, I consider how ever since Courtney and I broke up I've been sleeping around like I was twenty-two years old. Being single and being a bartender is like being a kid in a candy shop— all the flavors and colors available for the picking! It's thrilling. For the most part. Most of the time. But I'm fast discovering the downside. Like the other night when three women I'd slept with all found themselves at my bar at closing time. The fuck-off lights came on and none of them made a move to leave. I felt like a stud. Also weird and sleazy. Ideally they *all* would have been down with going back to my place to party, but that shit only flies when you're a celebrity or own a yacht in Ibiza.

I scribbled a note to the one I really liked to text me in fifteen, and passed it over the bar. Escorted the drunk one outside and waited for her taxi to come as she swayed against me, kissed her goodnight, and slipped the cabbie a twenty, giving him her address in Silver Lake. Back in the bar, I said a fast goodnight to the one talking to Devon and slipped out the back.

I felt *awesome.*

But the one I slipped the note to didn't text. I waited for what seemed like an appropriate amount of time—maybe she got lost?— and I texted her. But she didn't respond. I had gone from a foursome to a twosome to a nonesome and felt *terrible.*

"Here you go." The guy hands over my new set of keys as my phone buzzes.

Hey there stranger . . .—unsaved number.

I thank him and make a mental note to stop sleeping around.

In aisle 16, I find a one-horsepower air compressor I can use to blow accumulated dust off the refrigeration compressors so they don't overheat. Did you ever huff air duster? I did when I was sixteen. For about a week. One day in the library, I read I could also huff glue, wood stain, and paint, and then I read about the permanent damage they caused and I stopped.

In aisle 18, I grab needle-nose pliers for those hard-to-reach screws on the antique exit sign, which dates back to the 1940s. The bulbs they require to illuminate the thick, deep-red translucent glass are forty-watt and stay on all the time, but I love them so much that replacing the bulbs every couple of weeks feels worth it.

My phone buzzes.

I assume you saw last night's close out report re: Georgia, *so I set up an interview at 5. Okay that's it.*—Max

I set an alarm on my phone to remind me about the interview and cross my fingers it's a good one. Different people look for different things in their hires, but I'd rather hire a kook* than a startender any day—you know, the kind of bartender who wins flair contests. Who shakes pretty instead of effectively. Who orders obscure cocktails when visiting other bars, then reels off said cocktail's specs to the flummoxed bartender. While startenders may know their craft, their know-it-all attitude serves no one—not me, not my guests, and not, quite frankly, the startenders themselves. While it's wonderful to have amassed a wealth of knowledge, when it's wielded like a secret instead of a congenial, *Come with me on this journey!* attitude, well that sucks.

A great hire is someone with the desire to learn, even if they've worked in lots of other places. The key word is "collaboration." We hire personalities, not positions, because you can teach anyone service steps, but it's harder to get someone to care and be kind. I'm interested in people who have a thirst for learning and something

---

* "A term, most often used by aggro locals to describe surfers that pretend like they can surf when, in reality, they suck-ass."—Urban Dictionary

special to add to the mix. The ideal hospitality hire is someone who, should they stop by on their night off, can't help themselves from clearing a table sitting untended. The kind of person who's more than happy to relieve the host for a few should they need a break. This isn't a trait you can request in an interview or make protocol, but you can often sense if someone has it in them.

As I peruse aisle 20 for double-stick Velcro to tack to the ends of the BOH* curtain which always creeps off the wall and exposes the inventory we have stacked behind it, I think about how hiring is like dating: the person sitting across from you can be intriguing, but there's no proof things are going to work out. Questions I like to ask in interviews include: *Where are you from, why did you move here, and do you have any other aspirations? What excites and scares you about getting into the hospitality industry? How would you deal with an upset customer or co-worker? What are three things you would change about your community?* And then into speed dating: *What is your desert island cocktail? What did you want to be when you grew up? If I gave you $500 and two hours to spend it, what would you do?*

The truth is, the people who rock your world don't stay forever. The best ones are always going to leave, and they should—it's the natural evolution from someone's time as a disciple to their time as a teacher. David Rosoff hired me to open Osteria Mozza, and a year later I gave notice to open The Varnish. Devon Tarby was a Varnish regular who had never poured a drink in her life, got behind the stick, and is now a partner in Proprietors LLC. Kimmy Gatewood and Rebekka Johnson were Varnish servers who danced their way out of the bar and straight onto a billboard on Sunset Boulevard as stars of the TV show *GLOW*.

---

* Back-of-house.

There's the double-stick Velcro!

Buzz.

Yo E Dog, can you hold a table at The V for a date tonight @9:45? Want to give her the royal treatment.—Josh (friend of twenty-four years)

"The royal treatment" means making him look like a stud by sending over free drinks, and sitting down at their table to chat, being sure to include the perfect anecdote about something he did that's super cool.

My love for playing host is deeply ingrained in me, thanks to my parents, who have always been incredible hosts—cocktail parties, sit-down dinners, birthday bashes. Even a friend of mine dropping by wouldn't leave without being offered a Teisseire* mint syrup soda, a Petit-Beurre Lu,† or a piquant piece of Roquefort,‡ which wasn't relished by any kid, but if my mother could get *just one* hooked, she would consider it a victory.

My father's surprise fortieth birthday in our apartment on Twenty-third and First Avenue was my first bartending gig. I was seven years old, dressed in brown OshKosh B'gosh corduroy pants, a light blue button-up shirt, a navy sweater vest, and of course a bow tie. I didn't wear shoes, since we were at home. I did, however, have on striped socks.

While my dad's fraternity friends Steve Bauman and Ira Leitel kept him unaware of the preparations somewhere in town, I received a lesson in mixing drinks. The bar cart was a midcentury

---

* A French brand of flavored syrups.

† A rectangular shortbread cookie created in Nantes, France, in 1886.

‡ A blue ewe's-milk cheese from Occitanie. According to legend, a shepherd, enchanted by a shepherdess, left his cheese in a Combalou cave to give chase, and upon his return, discovered the cheese covered in mold. He tasted it and found it delicious.

modern Knoll creation with coated black metal shelves and brass legs. The top shelf ended at my head, and I could really only make out the colors of the tops of the bottles.

"Grab the dark brown bottle [sweet vermouth] and pour it to here on the glass," my mother instructed. "Then grab the brighter red bottle [Campari] and pour it into the same glass to here. Then the emerald-green bottle [Tanqueray gin] and finish at this mark," she pointed to an invisible line on the glass.

"Then add ice, and garnish it with an orange wedge from this bowl."

To this day, the Negroni is my go-to.

I text Josh back—*All the bells and whistles man*—then check out to a Muzak version of "Smells Like Teen Spirit" some psycho made and emerge from the bright white lights of Home Depot into the blinding yellow light of L.A., ready for my next action item.

CHAPTER SEVEN

SASHA

NEEDLESS TO SAY, THE MOMENT ANY PLEASURE IS TAKEN
AT DEMONSTRATING ONE'S SKILL AT SWORDSMANSHIP, ALL
POSSIBILITY OF TRUE SWORDSMANSHIP IS LOST.

—*Anonymous swordsman, sixteenth-century Japan*

It's May 20, 2004. My twenty-eighth birthday. Three friends and I look apprehensively around Eldridge Street: shuttered storefronts. Rats the size of cats. Old men in wifebeaters and black pleated slacks giving us the side eye. In the early 2000s, the Lower East Side wasn't riddled with crime like in the seventies, but it still had its fair share of mean streets.

I pull my Motorola StarTAC flip phone out of my pocket and call the secret number I wrested from a friend of a friend of a friend.

"Hello?" a woman's voice says.

"Yes, hi, my name is Eric. Eric Alperin." I walk a few paces down the street, away from my friends.

Silence.

"I don't have a reservation," I continue hurriedly, "but it's my birthday and I was hoping you had a table? For four?"

"Hold on," she says, placing her phone down. I hear tinkly music through the earpiece. See my friends on the corner standing in their Thursday finest: Stüssy, Abercrombie, one Von Dutch trucker hat. The phone abruptly disconnects and a sense of defeat lodges in my chest. I've been dying to check out Milk & Honey ever since hearing it was the anti-hip spot every hospitality industry member should go to. Accepting I'm going to have to wait a little longer, I look at my friends and shrug, which is when I notice, halfway down the block, an unmarked gray door cracking open and a woman peeking out. In movies, it's the big moments that make the protagonist feel like A Man—when he loses his virginity, when he wins a football game, when he gets married! And sure, I've done only one of those three things, so maybe I'm not the most objective observer, but in my experience it's the little things in life that send you, like the time I won the fifth-grade story-tell-a-thon by reciting Oscar Wilde's "The Nightingale and the Rose," when I worked out all the lyrics to "Jam on It," and right now, standing on Eldridge Street, being beckoned into Milk & Honey.

I wish I could tell you my friends and I were dazzled by the bar. That we were awed by the drinks and amazed by the service and stayed all night soaking up its every essence, but I would be lying. We ordered vodka martinis, spoke too loudly, stayed half an hour tops, then bounced to BOB Bar for nineties hip-hop.

But for weeks after our visit I couldn't stop thinking about the place: the stainless-steel box lying over a bed of crushed ice filled with tiny bowls holding fresh fruits and mint and dark red cherries that looked nothing like the neon processed stuff I was used to. There were towels in the bathroom! In the bars I'd been in, if they had towels at all, they were the abrasive brown variety pulled out of dispensers on the wall, but the ones at Milk & Honey were actual hand towels like the ones my parents had, stacked one on top of the other on a little shelf, folded side out. And I couldn't get the framed set of rules out of my head—a set of instructions on how to behave which I feared we broke, one after the other without meaning to.

We just didn't know any better.

Retroactively, I think my fascination with the place came from realizing that even though the bar was "fancy," none of us were belittled for wearing jeans and sneakers and a Von Dutch hat. We weren't made to feel like we ordered poorly. In our present-day mix-illogical revolution, it's not uncommon to encounter too-cool-for-school bartenders judging guests by their orders, giving good service if someone asks for a "classic" and virtually ignoring them if they request a 7&7. That night at Milk & Honey, the benevolent Christy Pope (who I would later train under at Little Branch) had no such attitude. She was generous and kind and served us the most delicious vodka martinis we'd ever tasted.

I BEGIN BADGERING EVERYONE I know for an introduction to the owner.

"Do you know Sasha Petraske?

"The guy who owns Milk & Honey?

"Can you give me his number?

"He doesn't have a number? Do you know when he works? Do you know where he lives?"

But nobody will tell me nothing. The guy's like a mob boss everyone's sworn to secrecy about.

So I give up.

Until months later, behind the stick at Lupa, I overhear two bartender friends, Wax and E-Rock from Von and BondST, discussing a new spot opening up on Leroy called Little Branch. A bigger version of Milk & Honey, they say. Same owner.

"What the fuck?" I exclaim. "I've been asking for weeks and you knew?"

"You didn't ask *me*," says E-Rock.

Wax eyeballs me and goes, "So you wanna become a mixologist, eh?"

"What? No. I don't even know what that means. Am I one?"

"Nope," Wax shakes his head. "Mixologists make their own syrups and have tons of cocktail books and house parties where they make 'concoctions' out of 'old books' for their 'friends.'"

"Really?" I feel like I miss everything. "Have you been to one?"

"Yep," they respond in unison.

"Why didn't you invite me?"

"We didn't know you wanted to be a *mixologist*," E-Rock grins.

"Fuck you man. I mean, I have a Mr. Boston guide and a ton of note cards from when I went to NBS, but—"

"What?" E-Rock interrupts. "If you went to *National Bartenders School*, how come you had no fucking clue how to make a Stoli-O and soda until you met me?"

I can see my GM waving drink tickets from the other end of the bar.

"I just want the phone number so I can get an interview," I tell them. "Is that too much to ask?"

"ERIC?" A GENTLEMAN SQUINTING INTO the daylight pokes his head out the front door of Little Branch. He has thick black hair and is wearing a wifebeater and boxers with socks pulled up to his knees.

"Mr. Petraske?" I ask.

"Pardon, but I just got up and am in a rush. Do come in while I get ready."

Just got up? I wonder as he ushers me down a set of darkened stairs to the basement bar, where I later learn he sleeps, living in a broom closet and using the men's bathroom to freshen up. The place has that vibe all bars have in their off-hours, like an old-timey boardwalk photo frozen in time that at any moment could come alive. The room smells sweet and dusky and briny and is humming with white noise.

"Let's sit over here," Sasha points to a chocolate leather booth. I slide into one side, he into the other. He begins to shine his shoes and asks me three questions—where I'm from, what I studied at school, and whether or not I get along with my mother.

After I answer he says, "I get to see my mother a lot. She lives nearby, and you will see her at the bar. Are you free to attend a staff meeting next week?"

SASHA WAS CENTRAL CASTING'S wet dream of the nutty professor. For most of the time I knew him, he slept in that broom closet at Little Branch and dressed in bespoke 1940s suits complete with suspenders and sock garters long before they became industry tropes. While his bars were great successes, Sasha was forever swimming in debt due to feeding his cats sushi-grade tuna, dining out nightly on shrimp cocktail and Champagne, dropping gobs on tailored suits, and caring for his mother, whom he loved most of all.

Sasha always spoke in a quiet, measured voice. He read *The Economist* every week from cover to cover, claiming it could teach you to bend spoons with your mind from across the room. To this day I have a subscription. Alas, I cannot bend spoons.

When Sasha was young, his grandmother gave him her *Communist Manifesto*, which informed his ideas of social justice and the notion that everyone was equal. Combined with an extensive use of psychedelics, this gave him a very singular way of interacting with the world. He was constantly deferential, holding doors open for people and allowing others to order before he did even when he was there first. It was as if he was in service even outside of the bar. Anyone who trained with Sasha will recall him saying, "You must take each shift as an opportunity to improve your craft, to lose your self-consciousness even though you are performing your tasks in front of an audience." But he lived his life like this—like it was a training ground for his work.

Sasha was also a microdoser, so you were never fully certain if his spaciness was a by-product of the high. Nibbles on mushrooms and torn edges of acid were quite normal for him. I think they helped him feel more at ease in the world. And if he happened to ingest a larger dose than planned, he'd ask one of his staff to hang out with him. Sasha surrounded himself with people who accepted him for who he was; no judgment. It was the same kind of ethos he extended to his bars and his staff.

But in L.A., Sasha seemed more relaxed, like he didn't have to be on point all the time. There was only one bar for him to focus on, and nobody knew him, versus the three bars he ran in New York and the hundreds of people clamoring for his attention. While our relationship started out as that of mentor/mentee, once we started working together on The Varnish, it changed and deepened. Even though Sasha's name carried cachet on the West Coast, he'd never opened a bar *not* by the seat of his pants, so we were learning together, and since he was in New York most of the time, he ceded much of the decision-making to me. In a very short time, we went from being mentor/mentee to being equals.

We were never as close as the month he slept on my couch before opening. Living with someone is very different from grabbing a drink together or even spending a weekend at an event in the same hotel room. Those four weeks we relied on each other 24/7, and while I would like to say I got to know him better, the truth is more nuanced. Sasha worked hard, but when we weren't working, he spent a lot of his own time reading and wandering who-knows-where. He was a gracious houseguest in the most compartmentalized of ways—eating all my cereal and using all my towels, but feeding Freeway, the cat. He'd never think of buying more cereal or doing the wash, but would come home with tins of Evanger's Organic Braised Chicken Dinner for Cats by the dozen.

In L.A., Sasha had a sense of peace he never seemed to have in Manhattan, and he was inspired to take better care of himself—more juicing, more yoga, longer hikes. And while he didn't come off as insecure, he would frequently express insecurity about his body. He wasn't in any way fat, just broad, and as much as he understood the benefits of being active and eating well, he was a natural-born hedonist, so he wasn't always successful at maintaining a balanced lifestyle. The first time we went to a public yoga class he left after a few minutes, completely freaked out about sweating and potentially passing gas in a roomful of elegant yogis while in Downward Dog. Unwilling to let that dash his hopes, he started streaming yoga classes. I'd be running errands and come back to my loft to find him in Warrior I, Bridge, or Savasana. And during bar training sessions at The Varnish, if he wasn't taking a nap in one of the booths during our breaks, he'd be in Downward Dog, shirt unbuttoned, sleeves rolled up, suspenders holding him all together.

"I find it a very good way to relax," he would tell whoever found him in this unexpected position. "I believe it prepares my brain for service."

When Sasha discovered a TED Talk by Amy Cuddy called "Your Body Language May Shape Who You Are," he ramped up his obsessions about his body and posture. In the talk, Cuddy says the way we carry our bodies affects not only those around us, but how we feel about ourselves. "Our bodies change our minds, our minds change our behavior, our behavior changes our outcome," says Cuddy.

As an actor, I'd spent more than my fair share of time getting in touch with my body, so Sasha's fascination with the correlation between the way we look and move and the way we think and feel was something I definitely got.

Sasha once told me he spent nine weeks training as a United States Army Ranger—a sector of Special Ops known for their strength, stamina, and preparedness to go into battle anytime, anywhere. The training is completed in three phases: the Crawl Phase, "designed to assess and develop the necessary physical and mental skills to complete combat missions"; the Walk Phase, which takes place in the mountains, where for twenty-one days, students learn to command and control a platoon-size patrol; and the Run Phase, which trains participants to operate effectively under conditions of extreme mental and physical stress.

I don't know why Sasha decided not to become a Ranger after all that work, but for sure the training had a serious effect on his physical and mental acuity—the way he appeared to be at rest while standing at the ready, his ability to stay alert no matter how little sleep he was running on, his aptitude for defusing challenging situations and remaining Buddha calm under pressure. Sasha's capacity for living in circumstances other people would find unmanageable—such as living in a bar, in a closet, without a shower (how he managed to stay so put together is beyond me)—was most certainly a result of all those weeks sleeping on the ground in the cold, being woken in the middle of the night to engage in simulated combat, and subsisting on two MREs* a day while enduring the

---

* Meals, ready to eat.

most grueling physical activities possible. It's also likely where Sasha learned that having good form was the result of a strong core and that good posture is not only beneficial to your physicality but communicates your intentions. In a bar, standing open to the floor with your body facing guests says, "Welcome to the party!" while slouching disconsolately, arms crossed, staring at your phone, tells everyone to go fuck themselves.

And so, Sasha trained staff on power poses to prepare them physically and mentally for a night of service. This idea that confident energy, via powerful posture, could help create a positive guest experience was a novel concept at the time, and when people were first training with Sasha, things could get a little weird.

"All right," he'd tell an assembled crew of bartenders, servers, barbacks, and hosts. "Let's do a power pose."

No one would know if he was joking or not. They'd look around at one another to gauge reactions. But Sasha was standing, so everyone would rise.

"Take two minutes and do your Wonder Woman posture," he'd demonstrate with feet apart, hands on hips, chin tilted upward. "There's a reason she stood like this," he'd say. "It made her feel powerful, prepared, ready to kick butt."

Titters inevitably followed as everyone spread their legs, posted their fists on their hips, and aimed their chins at the ceiling. But they would soon lapse into silence, captivated students, bedraggled warriors, ready for battle in the dim light of the daytime bar. It was always one of my favorite moments when staff was getting to know Sasha—I wish I had a time lapse of their expressions shifting across the hours, from thinking this was going to be a run-of-the-mill training session to not knowing what the hell they'd gotten themselves into to being a straight-up convert by day's end.

The indelible image of Sasha standing at attention holding

a coupe at its base is burned into many a cocktailian's mind—a power pose if ever there was one.

IN TRAINING SESSIONS, SASHA ALWAYS presented his ideas not as facts, or The Way We Do Things, but as concepts he found fascinating and wanted to share with others. He encouraged conversation so he wasn't the only one talking—(a) because he was inherently shy and (b) because he understood that everyone came to the table with something to offer.

"That's one way to do it," he would say if you executed something in a new or different way. Or "I bow to your superior logic" if you achieved something beyond what he imagined.

By creating an environment that offered people the chance to be unique contributors instead of cogs in a wheel, he built a bar in which every member who worked there felt purposeful. Places where no one merely towed the line but had a voice.

"We're allowed to try the drinks?" someone would inevitably ask when he instructed bartenders and servers to straw-test.

"Never trust a bartender that doesn't drink cocktails," he'd say with a smile. Then more seriously, "You surely can't recommend things to customers if you don't know what you're talking about. And while you may believe your specs are correct and wash lines[*] perfect, the only way to know whether your drinks are balanced is by tasting. If something is amiss, you must be able to identify the slightest variance in sweet and sour, then figure out what the cause

---

[*] The wash line is the level at which a finished cocktail rests inside the glass. It is also referred to as a drink's "collar." When a bartender mixes a cocktail to its appropriate temperature and dilution and pours it into a glass, it should leave nary a drop of liquid in the mixing vessel, and a perfect amount of space between the liquid and rim of the glass.

of those variances are. Was too much modifier jiggered? When was the citrus squeezed for service? Were the syrups correctly made? Is the drink too 'hot' because the booze was overpoured, or is it because it wasn't shaken long enough? You must taste your cocktails with a straw almost constantly to ensure consistency throughout the night."

Sasha's meditation on placement at the booths followed that of a restaurant, where there's a standard way to set the table—forks and napkins on the left, knives and spoons on the right. Glasses placed to the top right corner of the plate, and bread and salad plates to the left. Salt and pepper in the center. This gives a table symmetry and offers diners a sensible flow.

At The Varnish, the folded "clean" edge of a dental napkin[*] is placed in front of the guest a few inches from the table's edge. Drinks are placed dead center on the napkin. Garnishes are considered "peacock feathers" and are placed at two o'clock on the rim of the cocktail glass. This way you can see it as you sip, and your olfactory senses are flooded first with the herb or citrus or flower or ginger candy, second with the drink. If a garnish was placed at six o'clock, you'd get a mouthful of garnish; at three or nine o'clock it would hit the corner of your mouth. You'd also miss out on the visual fun and/or aromatics.

As for serving, Sasha would say, "The idea is to give the highest level of table service possible while minimizing interruption of the customer experience. You need to be as unobtrusive as possible. Rather than constantly interrupting the table's conversation, pass through their field of vision regularly, making yourself available to signals via eye contact or gesture. If you must reach

---

[*] We use Crosstex three-ply, white, uncoated dental bibs as napkins, which are exceptionally absorbent and durable.

past someone, softly say, 'Pardon my reach,' or whatever seems appropriate."

Sasha was a master at being intentional in every interaction—making eye contact with every person who approached the bar, especially if he wasn't able to serve them immediately. He knew just that simple gesture would let guests know they'd been seen and would soon be attended to. Asking open-ended, active questions such as "How are you, and have you been here before?" instead of "Can I get you a drink?" is a variation on the improv game "Yes, And?" Asking leading questions opens up the floor for someone to answer with more than just "Yes" or "No," which gives a bartender the opportunity to learn more about their guest.

"I'm great but I've never been here before" indicates they're in a good mood but need direction.

"I've had a terrible day and I'm so glad to be here" means they're gonna need a little extra love.

"Yeah I'm good, but I've been waiting for fucking ever and my date is *piiiiiissed*" is a signal for us to make something fast with a smile and find a clever way to turn their experience around.

WHEN SOMEONE IN THE INDUSTRY or a journalist used a term other than "bartender" to describe what we did, Sasha would gently explain, "We are not artists or 'mixologists'; just bartenders plying our trade by executing simple steps." And if someone got too heady about things, he'd wait until they finished their thought and say, "Bartending is not an intellectual pursuit. Which is not to say you shouldn't be intellectual, only that cocktails are meant to be experienced. If a guest would like to discuss ingredients or a drink's origins, by all means they should be indulged. But the pursuit is not to thrill with knowledge; it is to serve with consideration."

What Sasha understood best, and what he loved most, was how bars and everyone working within them had the opportunity to make people feel special. He was a true master at creating communities where the human condition could be nurtured and everyone felt comfortable sharing their joys, woes, and every emotion in between. It was never about what the place looked like or what style drinks they served; it was about the people who inhabited it. Sasha loved local bars maybe more than I do. There was definitely a Bukowski living inside him. In L.A. when he went missing, there were three places I'd look: if he wasn't at Cole's trying to unlock the science and lore behind their infamous French dip sandwiches with grizzled old-timers who had a lot to say on the subject, I'd head over to King Eddy's, where he drank 7&7s and talked to the frail, white-haired bartender. Lastly I'd check La Cita, where he drank cold Victorias while a mariachi band played live and the bartender was always trying to get him to do shots of tequila. All of these places offered a kind of service he believed in—genuine, kind, and unique.

And always, on his way out, if he didn't just sneak away without saying goodbye, he would pronounce, "May the Force be with you."

IN LATE 2006, WHEN SASHA and I first discussed the possibility of opening a bar together, he asked me a very simple question.

"Eric, do you like making lists?"

I laughed. "I don't know if 'like' is the word."

"Ah, well then," he said. "You'll be fine at opening a bar."

Lists are the obsessive-compulsives tether to a feeling of control. I have lists in my notebook. On my iPhone. Pushpinned to a corkboard propped against my living room floor and stuck to the walls of my office with myriad Post-its. Not only do I have written lists, but the photos folder on my phone is filled with pictures of glassware, menus, liquor bottles, receipts, a busted pipe, an HVAC filter, the combination to a lock, equipment serial numbers, pages of service manuals, cocktail recipes, pictures of handwritten lists, bar furniture and finishes, paint swatches, and wallpaper. And while this list-making practice doesn't guarantee everything in my day will go as planned, it at least gives me the illusion that with each ticking of a box, I get closer to the possibility of a successful day.

Reading another person's lists is incredibly intimate—on par with raiding their medicine cabinet or reading their diary—and while I don't advocate either, I leave you with a copy of one of Sasha's lists, a beautiful set of reminders from a man trying his best to live a healthy life and remain graceful in a world that often felt untenable.

I don't think he'd mind you seeing it.

1   Just because something isn't on your mind, doesn't mean you are not stressing about it

2   This checklist habit works

3   Large meal always = unhappy

4   Check messages when exiting subway

5   Eat before you're starving

6   Green juice

7   Cinnamon before carbohydrate

8   Remember things always take longer than you think they will

9   Handkerchief, comb, ID, cards, keys

10   Always put these in the same place

11   Don't put off dry cleaning/shoeshine

12   Remember cycling posture

13   Haircut every 14 days

14   If you say, "I will"→ Calendar

15   Turn ringer off in restaurants

16   Send reservation email ASAP

17   Remember you only dislike exercise if you're not warmed up

18   Allow adequate travel time

19   Being early is not a waste of time

20   Assume things go late

21   Speech is for communication

22   Remember reading posture

23   Listen, don't interrupt people

24   When you meet someone new, write down his or her name as soon as possible

25   Don't let this checklist habit go

CHAPTER EIGHT

METHODS AND RECIPES

I'm not going to pretend I've invented unique techniques and brought them to L.A. That an entire history didn't exist long before I showed up—places like Musso & Franks, the Tonga Hut and the Brown Derby, Tiki Ti, the Frolic Room, and Duke's. All Sasha and I did was recognize an opportunity to revitalize something we loved, in a city we hoped would embrace us.

There will be no extemporaneous history on cocktails or cocktail culture in this chapter; I leave that to David Wondrich. There will be no deep dives into shaking, stirring, and jiggering, because that has been eloquently covered by Dale DeGroff, Jim Meehan, Jeffrey Morgenthaler, and YouTube. No investigations into spirits, how they're distilled, or the way different flavor profiles and ABVs affect the balance of a drink will grace these pages, since F. Paul Pacult, Steve Olson, Doug Frost, and Thad Vogler have written definitively on these subjects. And finally, there will be no lessons on Superbags, rotovaping, or immersion circulators, as Alex Day and Dave Arnold are much cooler with those tricks than I am—and also, those techniques don't exist at The Varnish.

What I *do* want to do is drill down on the importance of code—a code championed at Milk & Honey and furthered by every member of its extended family. I want to clarify our doctrine for building drinks by the round, demystify the Mr. Potato Head method, hit on the basics we live by, and leave you with the 100-plus recipes a bartender heading into their first shift at The Varnish should know. So without further ado, we call to the stage . . .

MR. POTATO HEAD

Remember Mr. Potato Head? That beige plastic spud your grandmother bought you for your third birthday? He or she (there was also a Mrs. Potato Head) may have started off with red shoes, a mustache, and a black hat, but maybe you grew tired of that version and swapped out the black hat for a pink one, or paired your potato's tux with a sparkly tiara. It was crazy times!

Well, crazy as they were, today those potato-swapping antics apply to cocktails and, as coined by Phil Ward of Death & Co, are called the Mr. Potato Head method.

Here's how it works. Start with a classic recipe:

Daiquiri
2 ounces white rum
1 ounce lime juice
3/4 ounce simple syrup

By swapping out the rum for gin, voilà:

Gimlet
2 ounces gin
1 ounce lime juice
3/4 ounce simple syrup

Substitute the simple syrup with honey syrup and you get:

The Business
2 ounces gin
1 ounce lime juice
3/4 ounce honey syrup

Swap the gin back to rum with a little less lime juice, you got yourself a:

Honeysuckle
2 ounces white rum
3/4 ounce lime juice
3/4 ounce honey syrup

Switch out white rum for Jamaican rum, add two dashes of Angostura bitters, and there's your:

Brooklynite
2 ounces Jamaican rum
3/4 ounce lime juice
3/4 ounce honey syrup
2 dashes Angostura bitters

Replace the Jamaican rum with an equal-split spirit base of tequila and mezcal, an additional dash of bitters, and you are now the proud owner of the:

Oaxacanite
1 ounce blanco tequila
1 ounce mezcal
3/4 ounce lime juice

³/₄ ounce honey syrup
3 dashes Angostura bitters

These six cocktails are all based on the daiquiri and gimlet template. Plug and play away, but remember that you can't put ears in noses, which is to say you can't replace two ounces of rum with two ounces of lime juice and only one ounce of rum—you would be messing with the template and your drink would be unbalanced. The Mr. Potato Head method is ideal for coming up with new cocktails or simply categorizing and recalling variations on existing cocktails.

Here are some examples using martinis and manhattans, which are the same drink style—boozy and stirred up—using different base ingredients in the same three-ounce total proportion. They follow the same build as above, but by Mr. Potato Heading the modifiers (vermouths, liqueurs, and bitters) you can create new variations.

Martini
2 ounces gin
1 ounce dry vermouth

Add Benedictine and bitters, pull back on the dry vermouth, and you have a:

Poet's Dream
2 ounces gin
³/₄ ounce dry vermouth
¹/₄ ounce Benedictine
2 dashes orange bitters

Replace the dry vermouth and Benedictine with two different modifiers and you get the:

Deep Blue Sea
2 ounces gin
³/₄ ounce Cocchi Americano
¹/₄ ounce crème de violette (see p. 000)
2 dashes orange bitters

If you look back at the martini spec and swap out gin for rye, trade dry vermouth for sweet, and add some bitters, which are considered a seasoning, you have a:

Manhattan
2 ounces rye whiskey
1 ounce sweet vermouth
2 dashes Angostura bitters

Trade the one ounce of sweet vermouth for a total one ounce of dry vermouth, Picon, and Maraschino and there's your:

Brooklyn
2 ounces rye whiskey
¹/₂ ounce dry vermouth
¹/₄ ounce house Picon (see p. 000)
¹/₄ ounce Maraska Maraschino Cherry Liqueur

And for one more manhattan variation, let's bring back the sweet vermouth, some Cherry Heering, and a dash of absinthe for a:

Remember the Maine
2 ounces rye whiskey
³/₄ ounce sweet vermouth
¹/₄ ounce Cherry Heering
1 dash absinthe

All cocktails have some form of booze, modifying ingredients, and ice/water. Once you combine them, using varying techniques, you create different cocktail styles, of which there are five main ones.

- → OLD-FASHIONED/ON THE ROCK is boozy, built in the glass and served over a rock of ice.

- → MARTINIS AND MANHATTANS are stirred, silky, and served up.

- → THE SOUR involves shaking citrus, sugars, and traditionally egg white and is served up or on a rock.

- → THE HIGHBALL involves nonalcoholic mixers like citrus, juices, sugars, and always bubbles and is served long.

- → THE FIX involves citrus and sugars and is served over crushed or cracked ice.

I've added an additional section in the recipes for cream-based, dessert, flips (egg-yolk-based), and hot drinks that live just outside the main styles above.

Basically, a drink's "style" describes its dominant feature, the way it's prepared, and the way it's served. When guests don't know what they want to drink, a server might help by asking, "Would you like something shaken or stirred? Citrusy and refreshing? Sweet or herbaceous? Boozy and complex?" In this way, they're able to narrow down what style of cocktail they're inclined to drink.

There are also cocktail "branches," which are a way of organizing the different cocktails within those five styles. Each branch of cocktail has a parent cocktail. For instance, a gimlet is the matriarch and the Business is part of its family. The manhattan is the patriarch and the Brooklyn is a variation or offspring. You might hear a service bartender calling out to the personality bartender:

"Yo! What's in the Business again?"

"It's a gimlet, but with honey syrup!"

And the service bartender can easily Mr. Potato Head the honey syrup for simple syrup and thus create the Business.

A guest might say, "I loved your house martini. Can I get a variation?"

The server could offer them a Poet's Dream, which is a martini with a little Benedictine and orange bitters.

A cocktail server might yell across the well to the bartender, "What's in a London Buck again?"

And the bartender will likely reply, "Duh. It's a Moscow Mule with gin."

This allows the cocktail server to return to their table and explain that the drink is something the guest is familiar with, but using gin instead of vodka.

Michael Madrusan, a good friend and Milk & Honey veteran, spent years putting together an all-encompassing "branches" compendium, creating an organized way for bartenders to think about cocktail variations. If you live by the branches, recipes no longer feel like random numbers, which can make bartending seem overwhelming.

Here are a few more examples that illuminate Mr. Potato Head + styles + branches.

	OLD-FASHIONED	AMERICAN TRILOGY	SMUGGLER'S NOTCH
OLD-FASHIONED	2 ounces bourbon whiskey	1 ounce rye whiskey and 1 ounce applejack brandy	2 ounces aged rum
	3 dashes Angostura bitters	2 dashes orange bitters	2 dashes orange bitters
	1 white sugar cube	1 brown sugar cube	1 brown sugar cube
			Absinthe mist

ON THE ROCK	NEGRONI	MY OLD PAL	WHITE NEGRONI
	1 ounce gin	1 ounce rye whiskey	1½ ounces gin
	1 ounce Campari	1 ounce Campari	³/4 ounce Suze gentian liqueur
	1 ounce sweet vermouth	1 ounce sweet vermouth	³/4 ounce blanc vermouth

MARTINI	MARTINI	GIN & IT	FAIR & WARMER
	2 ounces gin	2½ ounces gin	2 ounces white rum
	1 ounce dry vermouth	½ ounce sweet vermouth	½ ounce sweet vermouth and ½ ounce curaçao

MANHATTAN	MANHATTAN	ROB ROY	GREENPOINT
	2 ounces rye whiskey	2 ounces Scotch whisky	2 ounces rye whiskey
	1 ounce sweet vermouth	1 ounce sweet vermouth	½ ounce yellow Chartreuse and ½ ounce sweet vermouth
	2 dashes Angostura bitters	2 dashes Angostura bitters	1 dash orange bitters and 1 dash Angostura bitters

	WHISKEY SOUR	PISCO SOUR	HARVEST SOUR
TRADITIONAL SOUR	2 ounces bourbon whiskey	2 ounces pisco	1 ounce rye whiskey and 1 ounce applejack brandy
	3/4 ounce lemon juice	3/4 ounce lemon juice	3/4 ounce lemon juice
	3/4 ounce simple syrup	3/4 ounce simple syrup	3/4 ounce simple syrup
	1 egg white	1 egg white	1 egg white
	Angostura bitters	Angostura bitters	Angostura bitters and Peychaud's bitters
		Cinnamon	Cinnamon

	FITZGERALD	BEE'S KNEES	GOLD RUSH
NONTRADITIONAL SOUR	2 ounces gin	2 ounces gin	2 ounces bourbon whiskey
	3/4 ounce lemon juice	3/4 ounce lemon juice	3/4 ounce lemon juice
	3/4 ounce simple syrup	3/4 ounce honey syrup	3/4 ounce honey syrup
	Angostura bitters		

	MOSCOW MULE	DARK 'N' STORMY	FLU COCKTAIL
HIGHBALL	2 ounces vodka	2 ounces Gosling's Black Seal rum	1 ounce rye whiskey and 1 ounce Cognac
	1/2 ounce lime juice	1/2 ounce lime juice	1/2 ounce lime juice
	3/4 ounce ginger syrup	3/4 ounce ginger syrup	3/4 ounce ginger syrup
	Top with soda water	Top with soda water	Top with soda water

	FIX	BRAMBLE	BRAZILIAN FIX
FIX	2 ounces spirit of choice	2 ounces gin	2 ounces cachaça
	3/4 ounce lemon juice	3/4 ounce lemon juice	3/4 ounce lime juice
	3/4 ounce simple syrup	3/4 ounce simple syrup	3/4 ounce honey syrup
		4 blackberries	1/4 ounce yellow Chartreuse

BUILDING DRINKS BY THE ROUND*

When a tableful of people orders drinks, chances are everyone will not order the same thing. How do you know what to make

---

* An edited and extended version of the original *Milk & Honey Service Manual*, written by Sasha Petraske and the staff of Milk & Honey for public domain, 1999–2009.

first? Does it matter? In the inimitable words of a writing teacher I know, "The answer is always YES!"

A cocktail is changeable. It is always in the process of getting warmer and, if it contains ice, more watery. Most customers will finish a drink that has become warm and watery, but they will not enjoy it as much. The goal is to deliver a drink to the customer in its coldest state, so they finish it while it's alive.* Since certain drinks decay much faster than others, texture- and temperature-wise, you must understand the durability of each style of cocktail in order to understand which might "die" sooner, as explained in chapter 5.

When you first start bartending, it will seem easier to build one drink at a time, but think through all repeating ingredients in a round and find where you can exercise economy of movement. If you're making two cocktails that have lime juice and simple syrup, you should reach for those ingredients only once. If two of the cocktails you're building require gin, you should be reaching for that bottle only once. And since we all fuck up sometimes, start building your drinks with the cheapest ingredients first and move up to the more expensive. If you screw up and have to toss your batch, you won't end up tossing the pricey booze, only the citrus and syrups.

When shooting for balanced cocktails, it's best to train yourself to use jiggers.† Once you become adept at hitting the measurements,

---

* When the legendary Harry Craddock, of the Savoy Hotel, was asked how quickly one should drink a cocktail, he replied, "Quickly, while it is still laughing at you." It was Sasha's belief that he was referring to the movement on the surface of a well-shaken, straight-up cocktail—if you shake hard enough, with cold enough ice, you will see bubbles forming and popping, and ice crystals swirling and shifting.

† Varying jiggers have measurement lines from 1/4 ounce all the way up to 2 1/2 ounces. In our recipes, we reference "scant," which is a little shy of what the measurement calls for, and "heavy," which means a hair's breadth more than what is listed. We also use a 3/8-ounce measurement that doesn't exist as a line on a jigger but means half of a 3/4-ounce measure.

you'll want to learn to maximize the use of your jiggers before discarding them. For example, you can overlap all citrus juices in the same jigger for a round of drinks. If there's three-quarters of an ounce of each citrus juice called for in a round of cocktails, you should start with lime, then lemon, then orange, then finish with grapefruit, all in the same three-quarter-ounce side of the jigger. I wouldn't reverse that order, though—the pulp of orange and grapefruit is larger and sweeter than lime or lemon. Make a fun game out of exhausting your jigger. Don't beat yourself up when you fuck up.

WITH SUGARS, START WITH SIMPLE SYRUP, then honey syrup, and end with ginger syrup, but again, not the other way around— ginger is the most complex of the three, and starting with it will cross-contaminate the purity of the honey or simple.

For any spirit served neat or on the rock, a fresh jigger is always used so that it's clean and pure, but when round-building cocktails, you can overlap accordingly: pouring any spirit into a jigger that has just measured vodka is always kosher. So if you're making three cocktails, and two of them have two ounces of gin each and the other one has two ounces of vodka, grab and pour the vodka first, followed by the gin.

Añejo/aged/dark rums are fine to follow in the same jigger after white rum, and agricole rum, thanks to its earthy and funky profile, is good to go after pouring añejo/aged/dark rums. It's also acceptable to pour bourbon after rye, and Scotch after bourbon. Gin and agave spirits, because of their strong flavor profiles, tend to end a jigger's life, requiring you to rinse it out before using it again. Lastly, be cautious of more complex modifiers like Benedictine, amari, Campari, Cointreau, absinthe, and Maraschino. Their flavors and textures are so unique and concentrated, they will contaminate the jigger, preventing any further overlapping of ingredients, unless that jigger is being used for the same cocktail.

If you can weave all this together for two cocktails, there's no reason why you can't take on a round of four, and soon six, moving the process along at a faster clip. It might seem aggressive for a home bartender, but this multitasking method will give you more hang time with your guests at your next party and less time stuck in the weeds.*

Below is the twelve-point sequencing to complete every possible style of drink.† It is a complex system based on simple, thoughtful, and deliberate moves, and illustrates how to set up your vessels, what to jigger first, what ice to use and when, and how to shake, stir, and serve.

1   Lay out the appropriate pieces of equipment for each drink on the work area, left to right, in the order you plan to build the drinks—for built-in-the-glass drinks, the appropriate glass; for shaken or stirred, the appropriate shaker‡ or mixing glass. This should also be the way the drinks are written on the order chit. Place all nonliquid modifying ingredients—sugar cubes, bitters,§ mint, limes—in the 18-ounce short side of the

---

* When you realize you can no longer keep up with service, i.e., "Oh fuck, I'm drowning!"

† Except hot drinks.

‡ When placing your tins down, put the 28-ounce tin behind the 18-ounce tin so the 18-ounce tin is closest to you. Joined together, these two tins create a seamless seal ideal for shaking with block ice. This is our house cocktail shaker.

§ Bitters can be tricky to manage in a traditional bottle. Not enough of a dash comes out when it's a new bottle, and too much splats out when the bottle is two-thirds empty. We employ the use of Japanese bitters bottles that are shaped with a long neck or barrel that allows for the same amount of bitters to be dashed out of the metal tip each time. Three dashes from a Japanese bitters bottle is equal to one good recipe dash.

shaking tin or mixing glass, or, in the case of an old-fashioned/ on-the-rock-style drink, directly in the drinking glass. Put egg whites or yolks in the larger 28-ounce side of the shaking tin so they don't interact with the citrus that will be poured into the shorter tin, until it's time to shake everything together.

2   Add citrus (no need to wash jigger between juices as long as you pour lime, then lemon, followed by orange, then grapefruit).

3   Add sweeteners/syrups: sugar, honey, ginger, orgeat.

4   Muddle by pressing down more than twisting. You want to wake up the ingredients, not turn them into a compost heap. Nothing needs to be muddled for more than two or three seconds. When muddling citrus wedges and chunks, you're "juicing with a muddler"—squeeze the juice out, then you're done.

5   Add more complex modifiers, such as vermouths, fortified wines, Campari, Cointreau, curaçao, Maraschino, Benedictine, crème de violette, amaro.

6   Add base alcohol. To review: spirits served on their own must be jiggered using a clean measure, but for a mixed drink involving citrus, there's no need to rinse the jigger between different rums or whiskeys. End with the strongest flavor. So jigger vodka, then white rum, then dark rum. Rinse your jigger or grab a new one and jigger gin or tequila. Then grab a new jigger and jigger rye, then bourbon, then Scotch.

7   Prepare any garnish cut to order that is not already in the mise en place, e.g., apple slices, twists, and any fruit requiring toothpicks for the rim of the glass.

8   Pull beer and white and rosé wines out of the cooler/refrigerator and pour into their appropriate glass on the tray. Pour bottled beer only two-thirds of the way. Place bottle next to glass. Pour red wine.

9   Now that *everything* that can be completed before icing is done, ice the most durable cocktails first, the most fragile last. Refer to chapter 5.

a)   Pour two-ingredient highballs (gin and tonic/rum and Coke) and spirits neat and on the rock.

b)   Ice your stirred, straight-up cocktails (martini/ manhattan). Use the spoon to lower the first medium-size piece of ice into the mixing glass to avoid splashing; the rest can be cracked in your hand into smaller pieces and shards and tossed in. Fill to the top with cracked ice. Start your stir, then leave the spoon in the glass. Stir intermittently as you go about building the rest of the round and make sure to plunge the ice at the top down into the mixing glass so that it hits the liquid. Add more cracked ice as needed if the liquid starts to dominate the ice.

c)   Ice your stirred-in-the-glass cocktails (old-fashioned/ negroni), using a spoon to lower the rock into the glass.

d)   For your shaken down/long drinks (Gold Rush/Tom Collins), remove double rocks and highball/collins glasses from glass froster or freezer. Shake with one rock of ice for your shaken and down. For highballs, shake with a smaller piece of ice, since the addition of soda water means they don't require as much dilution.

e)   Ice any "peasant"-style cocktails* (caipirinha/Gordon's Cup/smash) with cracked ice in the shorter eighteen-ounce tin, then seal the shaker with the larger twenty-eight-ounce tin and toss five times before dumping into

---

* In Brazilian Portuguese, *caipira* is slang to describe rural dwellers not used to big-city life, aka peasants. This evolved into referring to any drink made in the same style as a caipirinha, a "peasant cocktail."

a double rocks glass. Use a barspoon to "clean up" any muddled ingredients that look sloppy on the top by pushing them down.

f) For crushed-ice drinks (julep/mai tai/fix), dry-shake* and pour into their appropriate vessel, then top with crushed ice, which will provide all the meltage you need. For drinks with muddled mint, use a barspoon to clean up, pushing the mint to the lower third of the glass. Then use a barspoon or swizzle stick between your two hands, like a caveman trying to spark a fire, to integrate the crushed ice and ingredients and create a light frost on the sides of the glass.

g) For your shaken, straight-up cocktails (daiquiri/whiskey sour/French 75), remove glassware from froster and place on the work area or, if possible, directly on the serving tray. Dry-shake any cocktails with egg or cream to emulsify.† Add a single rock of ice and shake for temperature and dilution. Listen to your ice hurtling back and forth in the tin. Toward the end of the shake, you want to hear the single rock explode a little.

h) Immediately after your shake is done, you want to strain into a chilled glass. Always close the gate on the Hawthorne strainer completely, right up against the edge of the tin, so when you pour you see two even streams that flow like liquid fangs. This minimizes the size of the ice

---

* When you dry-shake a drink, you're shaking it without ice to mix the ingredients without adding dilution. When you're done, be careful when cracking open the shaker to add ice—ingredients like egg white expand when combined with other ingredients and may pop the tins apart.

† When you emulsify something, you mix two divergent liquids together, which thicken and become one. Like when you make a salad dressing of oil and vinegar, or in the case of a cocktail, you mix egg, citrus, and sugar to get a meringue-like/frothy texture.

crystals in the drink—if large slivers or chunks of ice are on top of the cocktail, the first taste will be of water; if there are no crystals at all, the cocktail will get warm sooner. Start the strain slowly and end with the tin inverted and empty by the end.*

i) Remove glassware for stirred-up drinks from the froster/freezer and strain cocktails using the julep strainer. Using the same strainer, scoop out any ice shards floating on top.

j) Float whipped cream† for White Russians, Dominicanas, etc.

10 Top off soda water in highballs, pour Champagne by the glass, top off any Champagne cocktails, and pour last third of bottled beer. Foam should be over the rim of the glass when it is handed to the guest.

11 Add garnishes and straws,‡ if not already done, at two o'clock on the rim of the glass. Make sure all drinks are evenly spaced on the tray, clockwise in the order they are listed on the chit, with a lit candle in the center.

12 Clean your work area, equipment, and bar, then check to see who needs your attention. Start next round.

---

* When shaking multiple straight-up cocktails, minimize the amount of time between each strain by partially shaking each drink, setting it down, shaking another partially, and then finishing off the first. If you're making more than three different straight-up cocktails, set the shakers down on ice. Try to strain off honey, pineapple-juice, and egg-white cocktails first, since they have a more forgiving structure, thanks to the combination of their ingredients and aeration. You must always "wake up" a shaker that you have set down before straining, which simply means to give it one more quick shake.

† To make whipped cream, add a few ounces of manufacturing/heavy whipping cream to an eight- to twelve-ounce squeeze bottle. Shake vigorously until frothy (about 20 seconds), then pour a thin layer over the cocktail in a clockwise fashion. The cream should float on the top.

‡ Preferably reusable.

Building drinks by the round allows a bartender to work "in the zone," dropping into an unconscious state and next thing you know, six hours have flown by. In the 1970s, Mihaly Csikszent-mihalyi, a psychologist at the University of Chicago, traveled the world conducting what would become one of the largest psychological surveys. He interviewed people of all creeds, colors, genders, and classes, and asked them when they were happiest and performed their best. The word that kept popping up was "flow," aka when your actions and decision-making process fall seamlessly into a fluid or "flowy" state of being.

Whatever kind of bartender you want to become—hotshot pro, home-based host, part-time slinger—I'd wager that if you practice building drinks by the round, one day soon, you'll find yourself quite unwittingly in the flow.

BARTENDER'S CHOICE

The Varnish is rooted in classic cocktails. Our menus feature a bouquet of six drinks—five representing a different style from one of the main branches, and the sixth is a Bartender's Choice. As noted in chapter 1, Bartender's Choice was created at Milk & Honey where there was no menu. Sasha said this was because he didn't want a cocktail list the size of a high school chemistry textbook overwhelming guests and gridlocking service. He told others he didn't make menus because he didn't have the money to print them up. The truth is likely a combination of the two.

What a guest receives as a Bartender's Choice is based not only on their responses to exploratory questions, but on what plays well with the rest of their table's order, or the orders of the people around them. It's not about wowing with some off-the-wall creation, but delivering just the right delicious drink in a timely manner, with a little variation and originality.

As explained in the Mr. Potato Head section, a manhattan lover would likely dig a Brooklyn variation. A staunch gimlet drinker might be down with the Business, and an old-fashioned loyalist would definitely drink a Monte Carlo or Fancy Free. Some guests can be exacting in their descriptions of what they want; others are open to being surprised and will try anything. Some people have no idea what they want or how to navigate the options. Finding out what people usually drink is a good baseline for making something that will either meet or exceed their expectations. We have a saying that frees up the minds of tentative or difficult guests: "If you don't like it, we'll drink it and make you something else."

Who can argue with that?

RECIPES

HOUSE-MADE STUFF

Developed by Sasha and refined over the years by the revolving staffs at Milk & Honey, Little Branch, and successive family bars, our syrups were created when we didn't have access to certain ingredients or when what was available didn't balance properly with our recipes.

Note: Any syrup that should be refrigerated has an approximate two-week shelf life. Those that are heavily booze-based are good almost indefinitely.

Nerd alert: For the sake of consistency we purchased a Brix meter (or refractometer) which measures the total dissolved solids in a drop of liquid by shining a light through it. For our purposes, the Brix meter quantifies the ratio of total dissolved sugars in any given syrup and calculates a percentage, which is the number you will find attached to these recipes. They're easy to use and you can pick one up online.

Specific brands listed are what we think work best. If you can't find them, by all means use what is available.

SIMPLE SYRUP—Brix 43%

In a sealable container, take 32 ounces (1 quart) hot water and add 3⅓ cups (750 grams) superfine sugar. (This recipe falls into the 1:1 category of sugar syrups, but ours is a little "weaker" than a 50:50 ratio, so it is a little less sweet than a traditional 1:1.)

Cover container and shake vigorously, transfer to a glass bottle, and store in refrigerator.

HONEY SYRUP—Brix 60%

Mix 3 parts orange blossom honey with 1 part hot water. Example: For 32 ounces (1 quart) syrup, combine 24 ounces honey and 8 ounces hot water. Stir thoroughly and store in a glass bottle in refrigerator.

GINGER SYRUP—Brix 43%

For every 32 ounces (1 quart) fresh ginger juice, add 3 cups (675 grams) superfine sugar. Stir until sugar is completely dissolved. Store in a glass bottle in refrigerator. For our ginger juice, we use fresh gingerroot and masticate it in a Breville Juice Fountain Elite 800JEXL.

CURAÇAO—Brix 40%

Combine 16 ounces Grand Marnier and 16 ounces simple syrup (equal parts). Stir thoroughly and store in a glass bottle.

GRENADINE—two methods—Brix 50%

1. Add 8 ounces pomegranate concentrate (we use Sadaf or FruitFast) to 24 ounces simple syrup. Stir thoroughly and store in a glass bottle in refrigerator. Make sure to use a concentrate, which has acidity, and not pomegranate molasses, which is already sweetened.

2. Reduce 32 ounces (1 quart) pomegranate juice by half in a saucepan over a low flame until it is about 16 ounces and then add 1 cup (225 grams) superfine sugar. Stir thoroughly and store in a glass bottle in refrigerator.

CRÈME DE VIOLETTE (Violette Liqueur)—Brix 34%

Combine 8 ounces Monin Violet Syrup, 8 ounces simple syrup, and 16 ounces spirit (vodka or gin). Stir thoroughly and store in a glass bottle.

FRUIT CUP (our house Pimm's)*—Brix 24%

Combine 12 ounces Beefeater London Dry Gin, 12 ounces Carpano Antica Formula Sweet Vermouth, 6 ounces Grand Marnier, and 4 ounces Cherry Heering liqueur. Stir thoroughly and store in a glass bottle.

HOUSE ORANGE BITTERS

Combine 7 ounces Bittermens Orange Cream Citrate and 3 ounces vodka. Stir thoroughly and store in a glass bottle.

HOUSE PICON—Brix 40%

Combine 24 ounces Bigallet China-China Amer, 6¾ ounces Suze gentian liqueur, and ½ cup (115 grams) superfine sugar. Stir thoroughly and store in a glass bottle.

---

* Pimm's is a British gin-based liqueur with bitter herbs and citrus, also known as a "fruit cup" because it's meant to be mixed in a long drink with an effervescent mixer and fruit garnish.

VARNISH ORGEAT—Brix 50%

Combine 5¹/₂ cups (44 ounces) simmering water with 2¹/₄ cups (315 grams) fresh raw almonds (skins on). Remove from heat but leave covered. Steep for 30 minutes; blend thoroughly; strain through a fine-mesh sieve* and press until the almond meal is dry; add enough water to resulting almond milk so that volume is 32 ounces (1 quart); be careful not to count the foam from the blending in this measurement. While still hot, add:

2 cups (450 grams) superfine sugar
2 cups (500 grams) raw cane (turbinado) sugar
1 teaspoon (6 grams) kosher salt
1¹/₄ ounces amaretto liqueur
1 ounce Cognac
20 drops rose water or to taste

Stir thoroughly until all sugar is dissolved and let cool uncovered. Store in a glass bottle in refrigerator.

LIME CORDIAL—Brix 40%

Place the peels of 12 limes into a sealable quart container; add 1¹/₂ cups (340 grams) superfine sugar, cover, and shake. Let sit 24 hours unrefrigerated. Add 16 ounces fresh lime juice and shake or stir to dissolve sugar completely; let sit 24 hours refrigerated. Strain and add 1¹/₂ ounces navy-strength gin to preserve. Stir thoroughly and store in a glass bottle in refrigerator.

---

* Preferably a chinois.

RASPBERRY SYRUP—Brix 43%

Lightly muddle 12 ounces raspberries with 16 ounces simple syrup. Let sit for 24 hours. Stir, then strain off raspberries and store in a glass bottle in refrigerator.

CAFÉ VARNISH—Brix 43%

Combine 13 ounces Bittermens New Orleans Coffee Liqueur, 17 ounces Caffé Lolita coffee liqueur, and ½ cup (115 grams) superfine sugar. Stir thoroughly and store in a glass bottle.

COCONUT CREAM—Brix 50%

Combine one 15-ounce can of Coco Lopez Cream of Coconut (shake well before opening), one 12-ounce can of Nature's Charm Evaporated Coconut Milk (shake well before opening), 1⅓ cups (300 grams) superfine sugar, and ¾ ounce 151 white rum to preserve. In a sealable container, shake all ingredients vigorously, transfer to a glass bottle, and store in refrigerator.

FRESH JUICES

Lime, lemon, orange, grapefruit . . . celebrate the pulp! In our opinion, it adds flavor and texture, which is the reason we never double-strain cocktails at The Varnish. I've had some tasty double-strained cocktails at other bars, but that's not how we do things. Do, though, remove any seeds or skins that sneak their way into the juice. Be careful if you use an electric juicer—when you hold the half-cut citrus on the reaming cone, don't let it grind too long or you'll get the bitterness of the pith in your juice. We have always used the tabletop Ra Chand J210 manual citrus juicer which allows us to squeeze with the correct pressure, and if necessary we can squeeze more à la minute during service if we run out without making too much noise.

We also make pineapple juice, which should be masticated in the juicer. Best to juice extra-ripe pineapples to capture more of their natural fruit sugars. Don't strain it, but let the juice settle and skim the foam off the top.

Technically, citrus juices should be used for only one night of service, then tossed at 2 a.m.

COCKTAILS

These 115 recipes are what new employees need to come armed with before their first solo shift at The Varnish. They represent a framework of cocktails that touch upon the relevant styles, branches, techniques, past menus, and common orders that occur at the bar. We are not a bar program of constant cocktail innovation. Creativity for us is a mindful meditation on how to execute cocktails and their ingredients consistently, ten thousand times over. The Varnish originals on this list came from tried-and-true R&D and follow the branch templates used in our bar family. Attributed originals from our friends and colleagues might vary slightly due to our in-house ingredients, but we proudly serve them. The Mr. Potato Head options listed beneath certain cocktail recipes are bonus and exemplify the wider range of options. There are specific brands listed for certain modifiers (amaris, vermouths, and liqueurs) because either the cocktail cannot be made without it or the selection is what we stand behind at The Varnish. You're welcome to experiment with any substitutes. Enjoy, catch a buzz, and if you get too drunk and pass out on your kitchen floor, no judgment.

Old-Fashioned and On the Rock—
Boozy and Built in the Glass

AFRICAN FLOWER

2 ounces (heavy) bourbon whiskey

1/4 ounce CioCiaro amaro

1/4 ounce crème de cacao

1 dash orange bitters

Glassware: Whiskey

Ice: Rock

Garnish: Orange twist

Method: Measure ingredients into a whiskey glass. Add ice rock and stir seven times. Express oils from an orange twist and then use as garnish.

Attribute: Becky McFalls, Little Branch

AMERICAN TRILOGY

1 ounce rye whiskey
1 ounce applejack brandy
1 brown sugar cube
2 dashes orange bitters
1 barspoon soda water
Glassware: Whiskey
Ice: Rock
Garnish: Orange twist
Method: Soak brown sugar cube with orange bitters in a whiskey glass. Add a barspoon of soda. Muddle into a paste. Measure rye whiskey and applejack brandy into the glass. Add ice rock and stir seven times. Express oils from an orange twist and then use as garnish.
Attribute: Richard Boccato and Micky McIlroy, Milk & Honey

Mr. Potato Head
Harvest Old-Fashioned: Sub one dash Angostura and one dash Peychaud's bitters for orange bitters and add both a lemon and an orange twist. (Sasha Petraske, Milk & Honey) *Note*: When two twists are used to garnish a drink, they are called "rabbit ears," in reference to how they are placed in the glass.

BETTER & BETTER

1½ ounces mezcal
½ ounce Smith & Cross Jamaican rum
¼ ounce Falernum
Glassware: Whiskey
Ice: Rock
Garnish: Lemon twist
Method: Measure all ingredients into a whiskey glass. Add ice rock and stir seven times. Express oils from a lemon twist and then use as garnish.
Attribute: Jan Warren, Dutch Kills

CHET BAKER

2 ounces aged rum

2 barspoons sweet vermouth

1 barspoon honey syrup

2 dashes Angostura bitters

Glassware: Whiskey

Ice: Rock

Garnish: Orange twist

Method: Measure all ingredients into a whiskey glass. Add ice rock and stir seven times. Express oils from an orange twist and then use as garnish.

Attribute: Sam Ross, Milk & Honey

DON LOCKWOOD

1 ounce Islay Scotch whisky

1 ounce bourbon whiskey

3/8 ounce maple syrup

1 dash Bittermens Xocolatl Mole Bitters

1 dash Angostura bitters

Glassware: Whiskey

Ice: Rock

Garnish: Orange twist

Method: Measure all ingredients into a whiskey glass. Add ice rock and stir seven times. Express oils from an orange twist and then use as garnish.

Attribute: Abraham Hawkins, Dutch Kills

FANCY FREE

2 ounces (heavy) bourbon whiskey

1/2 ounce (scant) Maraska Maraschino Cherry Liqueur

1 dash orange bitters

1 dash Angostura bitters

Glassware: Whiskey

Ice: Rock
Garnish: Orange twist
Method: Measure all liquid ingredients into a whiskey glass. Add ice rock and stir seven times. Express oils from an orange twist and then use as garnish.

IMPROVED WHISKEY COCKTAIL

2 ounces (heavy) rye whiskey
1/2 ounce (scant) Maraska Maraschino Cherry Liqueur
2 dashes Peychaud's bitters
2 dashes absinthe
Glassware: Whiskey
Ice: Rock
Garnish: Lemon twist
Method: Measure all liquid ingredients into a whiskey glass. Add ice rock and stir seven times. Express oils from a lemon twist and then use as garnish.

McKITTRICK

2 ounces (heavy) bourbon whiskey
1/2 ounce (scant) Pedro Ximénez sherry
2 dashes Bittermens Xocolatl Mole Bitters
Glassware: Whiskey
Ice: Rock
Garnish: Brandied cherry
Method: Measure all liquid ingredients into a whiskey glass. Add ice rock and stir seven times. Garnish with a brandied cherry.
Attribute: Theo Lieberman, Milk & Honey

Mr. Potato Head
Outlander: Sub Scotch whisky for bourbon whiskey and garnish with both a lemon twist and a brandied cherry. (Mikki Kristola, The Varnish)
Highlander: Sub Scotch whisky for bourbon whiskey, Cherry Heering for Pedro Ximénez sherry, remove mole bitters, and sub a lemon twist for brandied cherry. (Eric Alperin, The Varnish)

MONTE CARLO

---

2 ounces (heavy) rye whiskey
1/2 ounce (scant) Benedictine
2 dashes Angostura bitters
Glassware: Whiskey
Ice: Rock
Garnish: Lemon twist
Method: Measure all liquid ingredients into a whiskey glass. Add ice rock and stir seven times. Express oils from a lemon twist and then use as garnish.

Mr. Potato Head

Kentucky Colonel: Sub bourbon whiskey for rye whiskey.
Benedict Arnold: Sub Scotch whisky for rye whiskey.
Monte Carlos: Sub reposado tequila for rye whiskey. (Chris Bostick, Half Step)

NATIONAL VELVET

---

2 ounces (heavy) blended Scotch whisky
1/2 ounce (scant) Café Varnish (see p. 133)
2 dashes Bittermens Xocolatl Mole Bitters
Glassware: Whiskey
Ice: Rock
Garnish: Orange twist
Method: Measure all liquid ingredients into a whiskey glass. Add ice rock and stir seven times. Express oils from an orange twist and then use as garnish.
Attribute: Max Seaman, The Varnish

Mr. Potato Head

Revolver: Sub bourbon whiskey for Scotch whisky and orange bitters for mole bitters. (Jon Santer, Prizefighter)

NEGRONI

---

1 ounce gin

1 ounce Campari

1 ounce Carpano Antica Formula Sweet Vermouth

Glassware: Rocks

Ice: Rock

Garnish: Orange twist

Method: Measure all liquid ingredients into a rocks glass. Add ice rock and stir seven times. Express oils from an orange twist and then use as garnish.

Mr. Potato Head

My Old Pal: Sub rye whiskey for gin and sub lemon twist for orange twist.

Quill: Negroni with an absinthe rinse.

Famiglia Reale: 3/4 ounce Negroni with 1 ounce Champagne, garnish with a grapefruit twist. (Sam Ross, Attaboy)

Americano Cocktail: 1 1/2 ounces Campari and 1 1/2 ounces sweet vermouth without gin, served long over a spear, add soda water and an orange wedge garnish. *Note:* This option would now live in the highball family.

NICE LEGS

---

1 1/2 ounces gin

3/4 ounce Barolo Chinato

1/2 ounce Suze gentian liqueur

Glassware: Rocks

Ice: Rock

Garnish: Orange twist

Method: Measure all liquid ingredients into a rocks glass. Add ice rock and stir seven times. Express oils from an orange twist and then use as garnish.

Attribute: Chris Bostick, Half Step

Mr. Potato Head

Paige Ellis: Sub bourbon whiskey for gin. (Chris Bostick, Half Step)

OAXACA OLD-FASHIONED

1^{1}/$_{2}$ ounces reposado tequila

^{1}/$_{2}$ ounce mezcal

1 barspoon agave syrup

2 dashes Angostura bitters

Glassware: Whiskey

Ice: Rock

Garnish: Orange twist

Method: Measure all liquid ingredients into a whiskey glass. Add ice rock and stir seven times. Garnish with a flamed orange twist dropped into the drink.

Attribute: Phil Ward, Death & Co

OLD-FASHIONED

---

2 ounces bourbon whiskey

1 white sugar cube

3 dashes Angostura bitters

1 barspoon soda water

Glassware: Whiskey

Ice: Rock

Garnish: Orange twist

Method: Soak white sugar cube with Angostura bitters in a whiskey glass. Add a barspoon of soda water. Muddle into a paste, without completely dissolving; this helps integrate the sugar/bitters into the booze. Measure bourbon whiskey into the glass. Add ice rock and stir seven times. Express oils from an orange twist and then use as garnish.

POP QUIZ

---

2 ounces (heavy) bourbon whiskey

^{1}/$_{2}$ ounce (scant) Ramazzotti amaro

1 teaspoon simple syrup

2 dashes Bittermens Xocolatl Mole Bitters

Glassware: Whiskey

Ice: Rock

Garnish: Orange twist

Method: Measure all liquid ingredients into a whiskey glass. Add ice rock and stir seven times. Express oils from an orange twist and then use as garnish.

Attribute: Devon Tarby, The Varnish

SKIP JAMES

---

2 ounces brandy

2 barspoons Barolo Chinato

1 barspoon maple syrup

1 dash Angostura bitters

1 dash Peychaud's bitters

Glassware: Whiskey

Ice: Rock

Garnish: Lemon twist

Method: Measure all liquid ingredients into a whiskey glass. Add ice rock and stir seven times. Express oils from a lemon twist and then use as garnish.

Attribute: Bryan Tetorakis, The Varnish

SMUGGLER'S NOTCH

---

2 ounces aged rum

1 brown sugar cube

2 dashes orange bitters

1 barspoon soda water

Glassware: Whiskey

Ice: Rock

Garnish: Orange twist and absinthe mist

Method: Soak brown sugar cube with orange bitters in a whiskey glass. Add a barspoon of soda water. Muddle into a paste. Measure aged rum into the glass. Add ice rock and stir seven times. Express oils from an orange twist and then use as garnish. Finish off with a spritz of absinthe.

Attribute: Eric Alperin, The Varnish

TALENT SCOUT

2 ounces (heavy) bourbon whiskey

1/2 ounce (scant) curaçao

2 dashes Angostura bitters

Glassware: Whiskey

Ice: Rock

Garnish: Lemon twist

Method: Measure all ingredients into a whiskey glass. Add ice rock and stir seven times. Express oils from a lemon twist and then use as garnish.

Mr. Potato Head

Spanish Town: Sub aged rum for bourbon whiskey, no bitters, and sub grated nutmeg for lemon twist garnish.

Kentucky River: Sub crème de cacao for curaçao and peach bitters for Angostura bitters.

WHITE NEGRONI

1 1/2 ounces gin

3/4 ounce Dolin Blanc Vermouth

3/4 ounce Suze gentian liqueur

Glassware: Rocks

Ice: Rock

Garnish: Grapefruit twist

Method: Measure all ingredients into a rocks glass. Add ice rock and stir seven times. Express oils from a grapefruit twist and then use as garnish.

Attribute: Wayne Collins, created at Vinexpo, Bordeaux, France

Note: Originally created with Lillet Blanc, but at The Varnish we use a blanc vermouth.

Mr. Potato Head
Spring Blossom: Sub mezcal for gin and add 2 dashes of Bittermens Xocolatl Mole Bitters. (Gordon Bellaver, The Varnish)

Martinis and Manhattans—
Stirred Up and Silky

ARCHANGEL

2¼ ounces gin
¾ ounce Aperol
1 bruised cucumber slice
Glassware: Chilled coupe
Garnish: Lemon twist
Method: Lightly bruise a cucumber slice in the bottom of a chilled mixing glass. Measure remaining ingredients. Add cracked ice and stir. Strain into a chilled coupe. Express oils from a lemon twist and then use as garnish.
Attribute: Richard Boccato and Micky McIlroy, Little Branch

ASTORIA

---

2 ounces Dolin Dry Vermouth
1 ounce Old Tom gin
2 dashes orange bitters
Glassware: Chilled coupe
Garnish: Olive
Method: Measure all ingredients into a chilled mixing glass. Add cracked ice and stir. Strain into a chilled coupe. Garnish with an olive.
Note: This reverse vermouth/gin ratio is from *The Old Waldorf-Astoria Bar Book*.

ASTORIA BIANCO

2 ounces gin
1 ounce Dolin Blanc Vermouth
2 dashes orange bitters
Glassware: Chilled coupe
Garnish: Orange twist
Method: Measure all ingredients into a chilled mixing glass. Add cracked ice and stir. Strain into a chilled coupe. Express oils from an orange twist and then use as garnish.
Attribute: Jim Meehan, PDT

BAMBOO

1¹/₂ ounces Dolin Dry Vermouth
1¹/₂ ounces oloroso sherry
1 teaspoon Pedro Ximénez sherry
1 dash orange bitters
Glassware: Chilled coupe
Garnish: Orange twist
Method: Measure all ingredients into a chilled mixing glass. Add cracked ice and stir. Strain into a chilled coupe. Express oils from an orange twist and then use as garnish.
Note: Nowadays you see this drink made with fino sherry, which can be tough to keep fresh, so we started using an aged/oxidized oloroso and a Pedro Ximénez, which is a sweet cream sherry. Also, it is believed to have been historically made this way.

BOBBY BURNS

2 ounces Scotch whisky
¹/₂ ounce Cocchi Vermouth di Torino
¹/₄ ounce Benedictine
Glassware: Chilled coupe

Garnish: Shortbread cookie
Method: Measure all ingredients into a chilled mixing glass. Add cracked ice and stir. Strain into a chilled coupe. Garnish with a shortbread cookie on a side plate or, if possible, laid across the rim.

BOULEVARDIER

1½ ounces bourbon whiskey
¾ ounce Campari
¾ ounce Carpano Antica Formula Sweet Vermouth
Glassware: Chilled coupe
Garnish: None
Method: Measure all ingredients into a chilled mixing glass. Add cracked ice and stir. Strain into a chilled coupe.

Mr. Potato Head
Left Hand: Add Bittermens Xocolatl Mole Bitters and garnish with a brandied cherry. (Sam Ross, Attaboy)
Right Hand: Sub aged rum for bourbon whiskey, add Bittermens Xocolatl Mole Bitters, and garnish with an orange twist. (Micky McIlroy, Attaboy)

BROOKLYN

---

2 ounces rye whiskey
½ ounce Dolin Dry Vermouth
¼ ounce Amer Picon
¼ ounce Maraska Maraschino Cherry Liqueur
Glassware: Chilled coupe
Garnish: Brandied cherry
Method: Measure all ingredients into a chilled mixing glass. Add cracked ice and stir. Strain into a chilled coupe. Garnish with a brandied cherry.

Mr. Potato Head
Blue Collar: Sub sweet vermouth for dry vermouth and add 2 dashes of orange bitters and sub a lemon twist for brandied cherry. (Michael Madrusan, The Everleigh)

COLONIAL TIES

1 ounce rye whiskey
1 ounce Jamaican rum
1 brown sugar cube
2 dashes orange bitters
1 barspoon soda water
Absinthe
Glassware: Chilled whiskey
Garnish: Lemon twist
Method: Soak brown sugar cube with orange bitters in a chilled mixing glass. Add a barspoon of soda water and muddle into a paste. Add rye whiskey and Jamaican rum and stir with cracked ice. Strain into a chilled whiskey glass rinsed or atomized with absinthe. Express oils from a lemon twist and then use as garnish placed on the rim of the glass.
Attribute: Eric Alperin, The Varnish

DEEP BLUE SEA

2 ounces gin
³/₄ ounce Cocchi Americano
¹/₄ ounce crème de violette (see p. 131)
2 dashes orange bitters
Glassware: Chilled coupe
Garnish: Lemon twist
Method: Measure all ingredients into a chilled mixing glass. Add cracked ice and stir. Strain into a chilled coupe. Express oils from a lemon twist and then use as garnish.

FAIR & WARMER

2 ounces white rum
¹/₂ ounce Cocchi Vermouth di Torino
¹/₂ ounce curaçao
Glassware: Chilled coupe

Garnish: Lemon twist
Method: Measure all ingredients into a chilled mixing glass. Add cracked ice and stir. Strain into a chilled coupe. Express oils from a lemon twist and then use as garnish.

Mr. Potato Head

El Presidente: Sub dry vermouth for sweet vermouth, add a dash of grenadine, sub orange twist for lemon twist.

GREENPOINT

2 ounces rye whiskey
$^1/_2$ ounce yellow Chartreuse
$^1/_2$ ounce Cocchi Vermouth di Torino
1 dash orange bitters
1 dash Angostura bitters
Glassware: Chilled coupe
Garnish: Lemon twist
Method: Measure all ingredients into a chilled mixing glass. Add cracked ice and stir. Strain into a chilled coupe. Express oils from a lemon twist and then use as garnish.
Attribute: Micky McIlroy, Attaboy

MANHATTAN

2 ounces rye whiskey
1 ounce Cocchi Vermouth di Torino
2 dashes Angostura bitters
Glassware: Chilled coupe
Garnish: Brandied cherry
Method: Measure all ingredients into a chilled mixing glass. Add cracked ice and stir. Strain into a chilled coupe. Garnish with a brandied cherry.

Mr. Potato Head

Rob Roy: Sub Scotch whisky for rye whiskey.

Perfect Manhattan: Split sweet vermouth into equal parts sweet and dry vermouth. Sub lemon twist for brandied cherry.

MARTINI

2 ounces gin
1 ounce Dolin Dry Vermouth
Glassware: Chilled coupe
Garnish: Lemon twist or olive
Method: Measure all ingredients into a chilled mixing glass. Add cracked ice and stir. Strain into a chilled coupe. Garnish with either an olive or expressed lemon twist per guest request.
Note: Garnished with a pickled onion a Martini becomes a Gibson.

Mr. Potato Head

Amsterdam: Add orange bitters and garnish with lemon twist.
Gin & It: Sub 1/2 ounce sweet vermouth for dry vermouth, increase gin to 2 1/2 ounces, and express oils from a lemon twist and then use as garnish.
Dirty Martini: Split dry vermouth to equal parts dry vermouth and Dirty Sue olive brine.
Vodka Martini: 2 1/4 ounces vodka, 3/4 ounce dry vermouth.

POET'S DREAM

2 ounces gin
3/4 ounce Dolin Dry Vermouth
1/4 ounce Benedictine
2 dashes orange bitters
Glassware: Chilled coupe
Garnish: Lemon twist
Method: Measure all ingredients into a chilled mixing glass. Add cracked ice and stir. Strain into a chilled coupe. Express oils from a lemon twist and then use as garnish.

REMEMBER THE MAINE

2 ounces rye whiskey

3/4 ounce Cocchi Vermouth di Torino

1/4 ounce Cherry Heering

1 dash absinthe

Glassware: Chilled coupe

Garnish: Brandied cherry

Method: Measure all ingredients into a chilled mixing glass. Add cracked ice and stir. Strain into a chilled coupe. Garnish with a brandied cherry.

SAZERAC

2 ounces rye whiskey or Cognac (or equal parts, which is an example of a split-base cocktail)

1 white sugar cube

2 dashes Peychaud's bitters

1 barspoon soda water

Absinthe

Glassware: Chilled whiskey

Garnish: Lemon twist

Method: Soak white sugar cube with Peychaud's bitters in a chilled mixing glass. Add a barspoon of soda water and muddle into a paste. Add spirit of choice and stir with cracked ice. Strain into a chilled whiskey glass rinsed or atomized with absinthe. Express oils from a lemon twist and then use as garnish placed on the rim of the glass.

SKID ROW

2 ounces Bols Genever

1/2 ounce Ramazzotti amaro

1/2 ounce apricot liqueur (Rothman & Winter)

1 dash orange bitters

Glassware: Chilled coupe
Garnish: Orange twist
Method: Measure all ingredients into a chilled mixing glass. Add cracked ice and stir. Strain into a chilled coupe. Garnish with a flamed orange twist dropped into the drink.
Attribute: Eric Alperin, The Varnish

TRADITIONAL GIMLET

2 ounces gin
1 ounce lime cordial (see p. 132)
Glassware: Chilled coupe
Garnish: Lime twist
Method: Measure all ingredients into a chilled mixing glass. Add cracked ice and stir. Strain into a chilled coupe. Garnish with an unexpressed lime twist in the shape of a spiral on the rim.

Mr. Potato Head
Viva Villa: Sub blanco tequila for gin and garnish with sea salt.
Bennett: Add 2 dashes of Angostura bitters.

TUXEDO NO. 2

2 ounces gin
¾ ounce Dolin Dry Vermouth
¼ ounce Maraska Maraschino Cherry Liqueur
2 dashes orange bitters
Absinthe rinse
Glassware: Chilled coupe
Garnish: Lemon twist and brandied cherry
Method: Measure all ingredients (except absinthe) into a chilled mixing glass. Add cracked ice and stir. Strain into a chilled coupe rinsed or atomized with absinthe. Express oils from a lemon twist, then use as garnish and finish off with a brandied cherry.

The Sour—Shaken and Refreshing

⊰ DAIQUIRIS AND GIMLETS ⊱

THE BUSINESS

2 ounces gin
1 ounce lime juice
3/4 ounce honey syrup
Glassware: Chilled coupe
Garnish: Lime wedge
Method: Measure all ingredients into a cocktail shaker. Add a rock of ice and shake hard. Strain into a chilled coupe. Garnish with a lime wedge.
Attribute: Sasha Petraske, Milk & Honey

DAIQUIRI

2 ounces white rum
1 ounce (scant) lime juice
3/4 ounce simple syrup
Glassware: Chilled coupe
Garnish: None
Method: Measure all ingredients into a cocktail shaker. Add a rock of ice and shake hard. Strain into a chilled coupe.

Mr. Potato Head
Mint Daiquiri: Shake with 4–6 mint leaves and garnish with a mint leaf.
Strawberry Daiquiri: Shake with a muddled strawberry and garnish with strawberry.
Raspberry Daiquiri: Shake with 2–3 muddled raspberries or sub raspberry syrup (see p. 133) for simple syrup and garnish with a raspberry.
Captain's Blood: Add a dash of Angostura bitters and sub Jamaican rum for white rum and garnish with a lime wedge.

GIMLET (FRESH)

2 ounces gin
1 ounce (scant) lime juice
¾ ounce simple syrup
Glassware: Chilled coupe
Garnish: Lime wedge
Method: Measure all ingredients into a cocktail shaker. Add a rock of ice and shake hard. Strain into a chilled coupe. Garnish with a lime wedge.

Mr. Potato Head
Southside: Shake with 4–6 mint leaves and garnish with a mint leaf.
Eastside: Muddle and shake with 3 cucumber slices and 4–6 mint leaves, garnish with a cucumber slice.
Commodore: Sub bourbon whiskey for gin and add 2 dashes of orange bitters.

HEMINGWAY DAIQUIRI

1½ ounces white rum
¾ ounce Maraska Maraschino Cherry Liqueur
1 ounce grapefruit juice
½ ounce lime juice
Glassware: Chilled coupe
Garnish: Brandied cherry
Method: Measure all ingredients into a cocktail shaker. Add a rock of ice and shake hard. Strain into a chilled coupe. Garnish with a brandied cherry.

HIGH FIVE

1½ ounces gin
½ ounce Aperol
1 ounce grapefruit juice
½ ounce lime juice
½ ounce simple syrup

Glassware: Chilled coupe
Garnish: High five to your guest
Method: Measure all ingredients into a cocktail shaker. Add a rock of ice and shake hard. Strain into a chilled coupe. High fives all around!
Attribute: Alex Day, Death & Co

HONEYSUCKLE

2 ounces white rum
³/₄ ounce lime juice
³/₄ ounce honey syrup
Glassware: Chilled coupe
Garnish: None
Method: Measure all ingredients into a cocktail shaker. Add a rock of ice and shake hard. Strain into a chilled coupe.

Mr. Potato Head

Grandpa: Sub applejack brandy for white rum.
Brooklynite: Add 2 dashes of Angostura bitters and sub Jamaican rum for white rum.
Red Grasshopper: Sub blanco tequila for white rum and dust with cayenne powder. (Michael Madrusan, The Everleigh)

JACK ROSE

2 ounces applejack brandy
³/₄ ounce lime juice
³/₄ ounce grenadine
Glassware: Chilled coupe
Garnish: Apple slice
Method: Measure all ingredients into a cocktail shaker. Add a rock of ice and shake hard. Strain into a chilled coupe. Garnish with an apple slice.

Mr. Potato Head

Sky Pilot: Sub equal parts Jamaican rum and applejack brandy.

LAST WORD

1 ounce gin
³/₄ ounce green Chartreuse
³/₄ ounce Maraska Maraschino Cherry Liqueur
³/₄ ounce lime juice
Glassware: Chilled coupe
Garnish: Brandied cherry
Method: Measure all ingredients into a cocktail shaker. Add a rock of ice and shake hard. Strain into a chilled coupe. Garnish with a brandied cherry.

Mr. Potato Head
The Final Ward: Sub rye whiskey for gin, yellow Chartreuse for green Chartreuse, and lemon juice for lime juice. (Phil Ward, Death & Co)

OAXACANITE

1 ounce blanco tequila
1 ounce mezcal
³/₄ ounce lime juice
³/₄ ounce honey syrup
3 dashes Angostura bitters
Grapefruit twist
Glassware: Chilled coupe
Garnish: None
Method: Measure all ingredients into a cocktail shaker along with grapefruit twist. Add a rock of ice and shake hard. Strain into a chilled coupe.
Attribute: Ben Long, John Dory Oyster Bar
Note: Shaking a cocktail with a grapefruit twist in the tin is a technique we refer to as making it "regal."

OLD MAID

2 ounces gin
1 ounce lime juice

¾ ounce simple syrup

3 cucumber slices

4–6 mint leaves

Glassware: Chilled double rocks

Ice: Rock

Garnish: Cucumber and mint sprig

Method: Measure all ingredients into a cocktail shaker. Muddle cucumbers and mint. Add a rock of ice and shake hard. Strain into a chilled double rocks glass over a rock. Garnish with a mint sprig threaded through a cucumber slice on the rim.

Attribute: Sam Ross, Attaboy

Mr. Potato Head

Russian Maid: Sub vodka for gin.

Mexican Maid: Sub tequila for gin.

Kentucky Maid: Sub bourbon whiskey for gin.

French Maid: Sub Cognac for gin.

Scottish Maid: Sub Scotch whisky for gin.

LA OTRA PALABRA (THE OTHER WORD)

2 ounces mezcal

1 ounce lime juice

¼ ounce yellow Chartreuse

¼ ounce Maraska Maraschino Cherry Liqueur

1 barspoon agave syrup

Glassware: Chilled double rocks

Garnish: None

Method: Measure all ingredients into a cocktail shaker. Add a rock of ice and shake hard. Strain into a chilled double rocks glass.

Attribute: Eric Alperin, The Varnish

PINEAPPLE DAIQUIRI

2 ounces white or dark rum
¾ ounce pineapple juice
½ ounce lime juice
½ ounce simple syrup
Glassware: Chilled coupe
Garnish: None
Method: Measure all ingredients into a cocktail shaker. Add a rock of ice and shake hard. Strain into a chilled coupe.

SMOKE & MIRRORS

1 ounce blended Scotch whiskey
1 ounce Islay Scotch whisky
¾ ounce lime juice
¾ ounce simple syrup
3–5 mint leaves
Glassware: Chilled double rocks
Ice: Rock
Garnish: Mint sprig and Absinthe mist
Method: Measure all ingredients into a cocktail shaker. Muddle mint leaves. Add a rock of ice and shake hard. Strain into chilled double rocks glass. Garnish with a mint sprig and spray of absinthe.
Attribute: Alex Day, Death & Co

◄ NONTRADITIONAL—WITHOUT EGG WHITE ►

AVIATION

2 ounces gin
⅜ ounce crème de violette (see p. 131)
⅜ ounce Maraska Maraschino Cherry Liqueur

3/4 ounce lemon juice
Glassware: Chilled coupe
Garnish: Brandied cherry
Method: Measure all ingredients into a cocktail shaker. Add a rock of ice and shake hard. Strain into a chilled coupe. Garnish with a brandied cherry.

BEE'S KNEES

2 ounces gin
3/4 ounce lemon juice
3/4 ounce honey syrup
Glassware: Chilled coupe
Garnish: Lemon wedge
Method: Measure all ingredients into a cocktail shaker. Add a rock of ice and shake hard. Strain into a chilled coupe. Garnish with a lemon wedge.

CHAMPS-ÉLYSÉES

2 ounces Cognac
1/2 ounce green Chartreuse
3/4 ounce lemon juice
1/4 ounce simple syrup
Glassware: Chilled coupe
Garnish: Lemon twist
Method: Measure all ingredients into a cocktail shaker. Add a rock of ice and shake hard. Strain into a chilled coupe. Express oils from a lemon twist over the top and discard.

CORPSE REVIVER NO. 2

1 ounce gin
3/4 ounce Cointreau or Combier

³/₄ ounce Lillet or Cocchi Americano

³/₄ ounce lemon juice

1 dash absinthe

Glassware: Chilled coupe

Garnish: None

Method: Measure all ingredients into a cocktail shaker. Add a rock of ice and shake hard. Strain into a chilled coupe.

ENZONI

1 ounce gin

1 ounce Campari

³/₄ ounce lemon juice

³/₄ ounce simple syrup

6 Concord grapes

Glassware: Chilled double rocks

Ice: Rock

Garnish: Orange wedge and grape

Method: Measure all ingredients into a cocktail shaker. Muddle grapes. Add a rock of ice and shake hard. Strain into a chilled double rocks glass over an ice rock. Garnish with an orange wedge and a grape.

Attribute: Enzo Errico, Milk & Honey

Mr. Potato Head

Garibaldi: Sub rye whiskey for gin and 1 ounce lime juice for lemon juice. (Richard Boccato, Dutch Kills)

FITZGERALD

2 ounces gin

³/₄ ounce lemon juice

³/₄ ounce simple syrup

Glassware: Chilled coupe

Garnish: Angostura bitters

Method: Measure all ingredients into a cocktail shaker. Add a rock of ice and shake hard. Strain into a chilled coupe. Garnish with a dash of Angostura bitters.

Attribute: Dale DeGroff, Rainbow Room

Mr. Potato Head

Holland Razor Blade: Sub Bols Genever for gin and cayenne powder for Angostura bitters.

GOLD RUSH

2 ounces bourbon whiskey

3/4 ounce lemon juice

3/4 ounce honey syrup

Glassware: Chilled double rocks

Ice: Rock

Garnish: Lemon wedge

Method: Measure all ingredients into a cocktail shaker. Add a rock of ice and shake hard. Strain into a chilled double rocks glass over an ice rock. Garnish with a lemon wedge.

Attribute: T. J. Siegal, Milk & Honey

Mr. Potato Head

Golden Delicious: Sub applejack brandy for bourbon whiskey. (Jim Kearns, Milk & Honey)

Louisiana Purchase: Sub Cognac for bourbon whiskey. (Michael Madrusan, The Everleigh)

MARGARITA

1 1/2 ounces tequila

1 ounce Cointreau or Combier

3/4 ounce lime juice

Glassware: Chilled double rocks

Ice: Cracked

Garnish: Lime wedge; sea salt optional

Method: Measure all ingredients into a cocktail shaker. Add a rock of ice and shake hard. Strain into a chilled double rocks glass with cracked ice. Garnish with a lime wedge.

Note: If salt is requested, moisten half of a chilled glass's rim with a lime wedge and then roll the outside of the rim through a plate of sea salt. Clean up any salt that has made its way into the glass. Make sure to salt only half of the rim.

NAKED AND FAMOUS

1 ounce mezcal

³/₄ ounce Aperol

³/₄ ounce yellow Chartreuse

³/₄ ounce lime juice

Glassware: Chilled coupe

Garnish: None

Method: Measure all ingredients into a cocktail shaker. Add a rock of ice and shake hard. Strain into a chilled coupe.

Attribute: Joaquín Simó, Pouring Ribbons

PAPER PLANE

1 ounce bourbon whiskey

³/₄ ounce Aperol

³/₄ ounce Amaro Nonino

³/₄ ounce lemon juice

Glassware: Chilled coupe

Garnish: None

Method: Measure all ingredients into a cocktail shaker. Add a rock of ice and shake hard. Strain into a chilled coupe.

Attribute: Sam Ross, Attaboy

PENICILLIN

---

2 ounces Scotch whisky

3/8 ounce ginger syrup

3/8 ounce honey syrup

3/4 ounce lemon juice

1 barspoon Islay Scotch whisky (float)

Glassware: Chilled double rocks

Ice: Rock

Garnish: Candied ginger

Method: Measure all ingredients (except Islay Scotch whisky) into a cocktail shaker. Add a rock of ice and shake hard. Strain into a chilled double rocks glass over an ice rock. Float a barspoon of Islay Scotch whisky over the top. Garnish with a skewer of candied ginger.

Attribute: Sam Ross, Attaboy

Mr. Potato Head

Phoenix Down: Sub applejack brandy for Scotch whisky and sub a spray of absinthe for Islay float. (Daniel Eun, The Varnish)

SIDECAR

---

1 1/2 ounces Cognac

1 ounce Cointreau or Combier

1/2 ounce lemon juice

Glassware: Chilled coupe

Garnish: Lemon twist

Method: Measure all ingredients into a cocktail shaker. Add a rock of ice and shake hard. Strain into a chilled coupe. Express oils from a lemon twist and then use as garnish.

Mr. Potato Head

Chevalier: Add Angostura bitters.

Chelsea Sidecar: Sub gin for Cognac.

Royal Jubilee: Sub applejack brandy for Cognac.

Between the Sheets: 1 ounce Cognac, 1 ounce Cointreau or Combier, and 1 ounce lemon juice.

Brandy Crusta: Add ¹/₂ ounce Maraska Maraschino Cherry Liqueur, pull back to ¹/₂ ounce of Cointreau or Combier, and sugar half of the coupe's rim.

Southern Cross: Sub ³/₄ ounce aged rum and ³/₄ ounce Cognac for full 1¹/₂ ounces Cognac.

XYZ: Sub aged rum for Cognac.

SUGARPLUM

---

2 ounces gin

1 ounce grapefruit juice

¹/₂ ounce grenadine

1 teaspoon lemon juice

Glassware: Chilled coupe

Garnish: None

Method: Measure all ingredients into a cocktail shaker. Add a rock of ice and shake hard. Strain into a chilled coupe.

Attribute: Joseph Schwartz, Milk & Honey

Mr. Potato Head

Blinker: Sub rye whiskey for gin.

TOMMY'S MARGARITA

---

2 ounces tequila

1 ounce lime juice

¹/₂ ounce agave syrup

Glassware: Chilled double rocks

Ice: Cracked

Garnish: Lime wedge

Method: Measure all ingredients into a cocktail shaker. Add a rock of ice and shake hard. Strain into a chilled double rocks glass with cracked ice. Garnish with a lime wedge.

Attribute: Julio Bermejo, Tommy's Mexican Restaurant

VODKA DAISY

1¹/₂ ounces vodka
³/₄ ounce curaçao
³/₄ ounce lemon juice
1 splash soda water
Glassware: Chilled coupe
Garnish: Lemon twist
Method: Measure all ingredients (except soda water) into a cocktail shaker. Add a rock of ice and shake hard, for shorter duration than usual. Strain into a chilled coupe and top with a splash of soda. Express oils from a lemon twist over the top and discard.

Mr. Potato Head
Sub any spirit for vodka to get another Daisy variation.

WARD 8 NO. 2

---

2 ounces rye whiskey
¹/₂ ounce lemon juice
¹/₂ ounce orange juice
¹/₂ ounce grenadine
Glassware: Chilled coupe
Garnish: None
Method: Measure all ingredients into a cocktail shaker. Add a rock of ice and shake hard. Strain into a chilled coupe.

WATER LILY

---

1 ounce gin
³/₄ ounce Cointreau or Combier
³/₄ ounce crème de violette (see p. 131)
³/₄ ounce lemon juice

Glassware: Chilled coupe

Garnish: Orange twist

Method: Measure all ingredients into a cocktail shaker. Add a rock of ice and shake hard. Strain into a chilled coupe. Express oils from an orange twist over the top and discard.

Attribute: Richard Boccato, Dutch Kills

‑⧏ TRADITIONAL—WITH EGG WHITE ⧐‑

CHANTICLEER

2 ounces gin

¾ ounce lemon juice

¾ ounce simple syrup

3–4 raspberries

1 egg white

Glassware: Chilled sour

Garnish: Raspberry

Method: Muddle raspberries in a cocktail shaker. Add remaining ingredients and dry-shake. Add a rock of ice and shake hard. Strain into a chilled sour glass. Garnish with a raspberry.

CLOVER CLUB

2 ounces gin

½ ounce lemon juice

¾ ounce grenadine

1 egg white

Glassware: Chilled sour

Garnish: None

Method: Measure all ingredients into a cocktail shaker and dry-shake. Add a rock of ice and shake hard. Strain into a chilled sour glass.

HARVEST SOUR

1 ounce rye whiskey
1 ounce applejack brandy
3/4 ounce lemon juice
3/4 ounce simple syrup
1 egg white
Glassware: Chilled sour
Garnish: Angostura bitters, Peychaud's bitters, and grated cinnamon
Method: Measure all ingredients into a cocktail shaker and dry-shake. Add a rock of ice and shake hard. Strain into a chilled sour glass. Garnish with a design or streak of Angostura bitters and Peychaud's bitters over the top. Grate fresh cinnamon over the top.

PINK LADY

1 ounce gin
3/4 ounce applejack brandy
1/2 ounce lemon juice
3/4 ounce grenadine
1 egg white
Glassware: Chilled sour
Garnish: None
Method: Measure all ingredients into a cocktail shaker and dry-shake. Add a rock of ice and shake hard. Strain into a chilled sour glass.

PISCO SOUR

2 ounces pisco
3/4 ounce lemon juice
3/4 ounce simple syrup
1 egg white
Glassware: Chilled sour
Garnish: Angostura bitters and grated cinnamon

Method: Measure all ingredients into a cocktail shaker and dry-shake. Add a rock of ice and shake hard. Strain into a sour glass. Garnish with a design or streak of Angostura bitters over the top. Grate fresh cinnamon over the top.

WHISKEY SOUR

2 ounces bourbon whiskey
3/4 ounce lemon juice
3/4 ounce simple syrup
1 egg white
Glassware: Chilled sour
Garnish: Angostura bitters
Method: Measure all ingredients into a cocktail shaker and dry-shake. Add a rock of ice and shake hard. Strain into a sour glass. Garnish with a design or streak of Angostura bitters over the top.

Mr. Potato Head
New York Sour: Same as whiskey sour but strained over a rock in a chilled double rocks glass with a red wine float.

Highballs—Served Long with Bubbles

⊰ COLLINS ⊱

GRAPEFRUIT COLLINS

2 ounces gin
1 ounce grapefruit juice
1/2 ounce lemon juice
1/2 ounce simple syrup
2 dashes Peychaud's bitters
Soda water
Glassware: Chilled collins

Ice: Spear
Garnish: Lemon wedge and brandied cherry
Method: Measure all ingredients (except soda) into a cocktail shaker. Add a smaller rock of ice and shake for 5 seconds. Strain into a chilled collins glass with an ice spear. Top with soda and garnish with a lemon wedge and a brandied cherry.

TOM COLLINS

2 ounces gin
3/4 ounce lemon juice
3/4 ounce simple syrup
Soda water
Glassware: Chilled collins
Ice: Spear
Garnish: Lemon wedge and brandied cherry
Method Measure all ingredients (except soda) into a cocktail shaker. Add a smaller rock of ice and shake for 5 seconds. Strain into a chilled collins glass with an ice spear. Top with soda and garnish with a lemon wedge and a brandied cherry.

Mr. Potato Head
John Collins: Sub genever for gin.
Mike Collins: Sub Irish whiskey for gin.
Colonel Collins: Sub bourbon whiskey for gin.
Pierre Collins: Sub Cognac for gin.
Sandy Collins: Sub Scotch whisky for gin.
Stay Up Late: Sub 1/2 ounce Cognac and 1 1/2 ounces gin for full 2 ounces gin, and sub orange wedge for lemon wedge.

◄ RICKEYS ►

GIN RICKEY

2 ounces gin
1 ounce lime juice
3/4 ounce simple syrup
Soda water
Glassware: Chilled collins
Ice: Spear
Garnish: Lime wedge
Method: Measure all ingredients (except soda) into a cocktail shaker. Add a smaller rock of ice and shake for 5 seconds. Strain into a chilled collins glass with an ice spear. Top with soda and garnish with a lime wedge.

Mr. Potato Head
Southside Rickey: Shake with 4–6 mint leaves and sub a mint sprig garnish for lime wedge.
Eastside Rickey: Southside Rickey with muddled cucumber and garnished with a mint sprig and cucumber.
Tritter Rickey: Southside Rickey with 3 dashes of absinthe. (Michael Tritter, Milk & Honey)
Ivy Fizz: Southside Rickey but sub vodka for gin.

MEXICAN FIRING SQUAD

2 ounces tequila
3/4 ounce lime juice
3/4 ounce grenadine
2 dashes Angostura bitters
Soda water
Glassware: Chilled collins
Ice: Spear
Garnish: Orange wedge and brandied cherry

Method: Measure all ingredients (except soda) into a cocktail shaker. Add a smaller rock of ice and shake for 5 seconds. Strain into a chilled collins glass with an ice spear. Top with soda and garnish with an orange wedge and a brandied cherry.

LA PALOMA (FRESH)

2 ounces blanco tequila
1 ounce grapefruit juice
1/2 ounce lime juice
1/2 ounce simple syrup
Soda water
Glassware: Chilled collins
Ice: Spear
Garnish: Grapefruit wedge and sea salt
Method: Measure all ingredients (except soda) into a cocktail shaker. Add a smaller rock of ice and shake for 5 seconds. Strain into a chilled collins glass with an ice spear. Top with soda, garnish with a grapefruit wedge, and sprinkle sea salt on top.

ROME WITH A VIEW

1 ounce Campari
1 ounce Dolin Dry Vermouth
1 ounce lime juice
3/4 ounce simple syrup
Soda water
Glassware: Chilled collins
Ice: Spear
Garnish: Orange wedge
Method: Measure all ingredients (except soda) into a cocktail shaker. Add a smaller rock of ice and shake for 5 seconds. Strain into a chilled collins glass with an ice spear. Top with soda and garnish with an orange wedge.
Attribute: Micky McIlroy, Attaboy

VATICAN CITY

1 ounce Suze gentian liqueur
1 ounce Dolin Blanc Vermouth
1 ounce lime juice
3/4 ounce simple syrup
Soda water
Glassware: Chilled collins
Ice: Spear
Garnish: Grapefruit twist
Method: Measure all ingredients (except soda) into a cocktail shaker. Add a smaller rock of ice and shake for 5 seconds. Strain into a chilled collins glass with an ice spear. Top with soda and express oils from a grapefruit twist, then use as garnish.
Attribute: Mikki Kristola, The Varnish

⊰ BUCKS ⊱

FLORADORA

2 ounces gin
1/2 ounce lime juice
1/2 ounce ginger syrup
1/2 ounce simple syrup (or sub raspberry syrup [see p. 133] for simple syrup and forgo the fresh raspberries)
3–4 raspberries
Soda water
Glassware: Chilled collins
Ice: Spear
Garnish: Ginger candy and raspberry
Method: Measure all ingredients (except soda) into a cocktail shaker. Add a smaller rock of ice and shake for 5 seconds. Strain into a chilled collins glass with an ice spear. Top with soda and garnish with a skewered ginger candy and a raspberry.

GIN-GIN MULE

2 ounces gin

1 ounce lime juice

$1/2$ ounce ginger syrup

$1/2$ ounce simple syrup

Mint

Soda water

Glassware: Chilled collins

Ice: Spear

Garnish: Ginger candy and mint sprig

Method: Measure all ingredients (except soda) into a cocktail shaker. Add a smaller rock of ice and shake for 5 seconds. Strain into a chilled collins glass with an ice spear. Top with soda and garnish with a skewered ginger candy and a mint sprig.

Attribute: Audrey Saunders, Pegu Club

MOSCOW MULE

2 ounces vodka

$3/4$ ounce ginger syrup

$1/2$ ounce lime juice

Soda water

Glassware: Chilled collins

Ice: Spear

Garnish: Ginger candy

Method: Measure all ingredients (except soda) into a cocktail shaker. Add a smaller rock of ice and shake for 5 seconds. Strain into a chilled collins glass with an ice spear. Top with soda and garnish with a skewered ginger candy.

Mr. Potato Head

Palma Fizz: Add a spray of rose water. (Sasha Petraske, Milk & Honey)

London Buck: Sub gin for vodka.

Dark 'n' Stormy: Sub Gosling's Black Seal rum for vodka.

Cablegram: Sub rye whiskey for vodka.

Flu Cocktail: Sub equal parts rye whiskey and Cognac for vodka.

El Diablo: Sub blanco tequila for vodka and finish off with a crème de cassis float.

NEW ORLEANS BUCK

2 ounces aged rum
1 ounce orange juice
½ ounce lemon juice
½ ounce ginger syrup
2 dashes Angostura bitters
Soda water
Glassware: Chilled collins
Ice: Spear
Garnish: Ginger candy and orange wedge
Method: Measure all ingredients (except soda) into a cocktail shaker. Add a smaller rock of ice and shake for 5 seconds. Strain into a chilled collins glass with an ice spear. Top with soda and garnish with a skewered ginger candy and an orange wedge.

RANGOON

2 ounces Pimm's or fruit cup (see page 131)
¾ ounce ginger syrup
Lime wedge
Lemon wedge
Blackberry or raspberry
2 cucumber slices
Sprite
Glassware: Chilled collins
Ice: Spear
Garnish: 3 Cucumber slices
Method: Muddle berries, citrus, and cucumber in a chilled collins glass without ice. Add Pimm's and ginger syrup. Add ice spear and stir several times. Top with Sprite. Garnish with 3 cucumber slices.

⇥ TRADITIONAL FIZZES—WITH EGG WHITE ⇤

GIN FIZZ

2 ounces gin
3/4 ounce lemon juice
3/4 ounce simple syrup
1 egg white
Soda water
Glassware: Chilled fizz
Garnish: None
Method: Measure all ingredients (except soda) into a cocktail shaker and dry-shake. Add a rock of ice and shake hard. Strain into a chilled fizz glass. Top with soda.

Mr. Potato Head
Alabama Fizz: Shake with 4–6 mint leaves and garnish with a mint leaf.

RAMOS GIN FIZZ

11/2 ounces gin
3/8 ounce lemon juice
3/8 ounce lime juice
1/2 ounce simple syrup
1/2 ounce heavy cream
1 egg white
5 dashes orange flower water
Soda water
Glassware: Chilled fizz
Garnish: Orange twist
Method: Measure all ingredients (except soda) into a cocktail shaker and dry-shake. Add 2 ounces pebble ice and shake until completely dissolved. Strain into a chilled fizz glass. Top with soda. Express oils from an orange twist and then use as garnish. Serve with a straw.

Note: There is an ongoing debate about how long one must shake a Ramos to allow the egg and cream to froth properly. This is the only cocktail we shake with pebble ice in order to achieve the desired texture.

Mr. Potato Head
Southern Whiskey Fizz: Sub applejack brandy for gin and remove orange flower water.

SILVER FOX

1¹/₂ ounces gin
³/₄ ounce lemon juice
¹/₂ ounce orgeat (see p. 132)
1 egg white
Soda water
¹/₂ ounce amaretto (float)
Glassware: Chilled fizz
Garnish: None
Method: Measure all ingredients into a cocktail shaker (except soda and amaretto) and dry-shake. Add a rock of ice and shake hard. Strain into a chilled fizz glass. Top with soda and float amaretto.
Attribute: Richard Boccato, Milk & Honey

SILVER LINING

1¹/₂ ounces rye whiskey
³/₄ ounce Licor 43
³/₄ ounce lemon juice
1 egg white
Soda water
Glassware: Chilled collins
Garnish: None
Method: Measure all ingredients (except soda) into a cocktail shaker and dry-shake. Add a rock of ice and shake hard. Strain into a chilled collins glass with ice spear. Top with soda.
Attribute: Joseph Schwartz, Milk & Honey

⊰ CHAMPAGNE ⊱

AIRMAIL

1 ounce light rum
$1/2$ ounce lime juice
$1/2$ ounce honey syrup
Champagne
Glassware: Chilled Champagne flute
Garnish: None
Method: Measure all ingredients (except Champagne) into a cocktail shaker. Shake ingredients together for 5 seconds with a smaller rock of ice and strain into a chilled Champagne flute. Top with Champagne.

Mr. Potato Head
Blackmail: Add a muddled blackberry. (Little Branch)

FRENCH 75

1 ounce gin or Cognac
$1/2$ ounce lemon juice
$1/2$ ounce simple syrup
Champagne
Glassware: Chilled Champagne flute
Garnish: Lemon twist
Method: Measure all ingredients (except Champagne) into a cocktail shaker. Shake ingredients together for 5 seconds with a smaller rock of ice and strain into a chilled Champagne flute. Top with Champagne and express oils from a lemon twist and then use as garnish.

HARRY'S PICK-ME-UP

1 ounce Cognac
$^1/_2$ ounce lemon juice
$^1/_2$ ounce grenadine
Champagne
Glassware: Chilled Champagne flute
Garnish: Lemon twist
Method: Measure all ingredients (except Champagne) into a cocktail shaker. Shake ingredients together for 5 seconds with a smaller rock of ice and strain into a chilled Champagne flute. Top with Champagne and express oils from a lemon twist and then use as garnish.

The Fix—Crushed and Cracked Ice

◄ PEASANTS AND SMASHES ►

CAIPIRINHA

2 ounces cachaça
$^3/_4$ ounce simple syrup
6 lime quarters
1 white sugar cube
Glassware: Chilled double rocks
Ice: Cracked
Garnish: None
Method: Muddle together in a cocktail shaker the lime quarters, sugar cube, and simple syrup. Add cachaça and cracked ice. Shake five times and dump into a chilled double rocks glass.

Mr. Potato Head
Brazilian Mojito: Add muddled mint.
Caipiroska: Sub vodka for cachaça.

GORDON'S CUP

2 ounces gin

3/4 ounce simple syrup

5 lime quarters

4 cucumber slices

Glassware: Chilled double rocks

Ice: Cracked

Garnish: Sea salt

Method: Muddle together in a cocktail shaker the lime quarters, cucumber slices, and simple syrup. Add gin and cracked ice. Shake five times and dump into a chilled double rocks glass. Garnish with a pinch of sea salt.

Attribute: Sasha Petraske, Milk & Honey

Mr. Potato Head

Gordon's Breakfast: Add 2–3 dashes of Cholula Hot Sauce and 3 drops of Worcestershire sauce.

El Guapo: Add 2–3 dashes of Cholula Hot Sauce and sub tequila for gin. (Sam Ross, Attaboy)

WHISKEY SMASH

2 ounces bourbon whiskey

3/4 ounce simple syrup

3–4 lemon quarters

4–6 mint leaves

Glassware: Chilled double rocks

Ice: Cracked

Garnish: Lemon wedge and mint sprigs

Method: Muddle together in a cocktail shaker the mint, lemon quarters, and simple syrup. Add bourbon whiskey and cracked ice. Shake and strain into a chilled double rocks glass over fresh cracked ice. Garnish with a lemon wedge and mint sprigs.

Attribute: Dale DeGroff, Rainbow Room

Note: This drink can also be strained over crushed ice.

Mr. Potato Head

Sub any spirit for bourbon whiskey to get another smash variation.

◅ JULEPS, COBBLERS, AND SWIZZLES ►

BRAMBLE

2 ounces gin

3/4 ounce lemon juice

3/4 ounce simple syrup

4 blackberries

Glassware: Chilled double rocks

Ice: Pebble

Garnish: Blackberry

Method: Muddle together in a cocktail shaker the blackberries, lemon juice, and simple syrup. Add gin. Dry-shake all ingredients and dump into a chilled double rocks glass. Top with pebble ice and lightly swizzle. Garnish with a blackberry. Add topping of pebble ice.

Attribute: Dick Bradsell, London

Mr. Potato Head

Rumble: Sub white rum for gin and sub raspberries for blackberries.

BRAZILIAN FIX

2 ounces cachaça

3/4 ounce lime juice

3/4 ounce honey syrup

1/4 ounce yellow Chartreuse (float)

Glassware: Chilled double rocks

Ice: Pebble

Garnish: Mint sprigs

Method: Measure all ingredients (except yellow Chartreuse) into a cocktail shaker. Dry-shake and dump into a chilled double rocks glass. Top with

pebble ice and swizzle. Float yellow Chartreuse. Garnish with mint sprigs. Add topping of pebble ice.
Attribute: Eric Alperin, The Varnish

FIX

2 ounces spirit of choice
³/₄ ounce lemon juice
³/₄ ounce simple syrup
Glassware: Chilled double rocks
Ice: Pebble
Garnish: Lemon wedge
Method: Dry-shake all ingredients in a cocktail shaker and dump into a chilled double rocks glass. Top with pebble ice and swizzle. Garnish with a lemon wedge. Add topping of pebble ice.

GEORGIA JULEP

2 ounces Cognac
¹/₂ ounce peach liqueur
1 brown sugar cube
6–8 mint leaves
Glassware: Julep tin
Ice: Pebble
Garnish: Mint sprigs and powdered sugar
Method: Muddle the brown sugar cube, mint, and peach liqueur together in a julep tin. Add Cognac. Top with pebble ice and swizzle. Garnish with mint sprigs and powdered sugar. Add topping of pebble ice.

MAI TAI

1 ounce Coruba dark rum
1 ounce Appleton Estate V/X aged rum

¾ ounce lime juice
½ ounce (heavy) orgeat (see p. 132)
½ ounce (scant) curaçao
Glassware: Chilled double rocks
Ice: Pebble
Garnish: Mint sprigs
Method: Dry-shake all ingredients in a cocktail shaker and dump into a chilled double rocks glass. Top with pebble ice and swizzle. Garnish with mint sprigs. Add topping of pebble ice.

MINT JULEP

2 ounces bourbon whiskey
¼ ounce simple syrup
1 white sugar cube
6–8 mint leaves
Glassware: Julep tin
Ice: Pebble
Garnish: Mint sprigs and powdered sugar
Method: Muddle the white sugar cube, mint, and simple syrup together in a julep tin. Add bourbon whiskey. Top with pebble ice and swizzle. Garnish with mint sprigs and powdered sugar. Add topping of pebble ice.

Mr. Potato Head
Prescription Julep: Sub ½ ounce rye whiskey and 1½ ounces Cognac for full 2 ounces bourbon whiskey.

MOJITO

2 ounces white rum
1 ounce lime juice
¾ ounce simple syrup
1 brown sugar cube
8–10 mint leaves

Glassware: Chilled double rocks
Ice: Pebble
Garnish: Mint sprigs
Method: Gently muddle the brown sugar cube, mint, lime juice, and simple syrup together in a cocktail shaker. Add white rum. Dry-shake and dump into a chilled double rocks glass. Top with pebble ice and swizzle. Garnish with mint sprigs. Add topping of pebble ice.

'OUMUAMUA (*OH-MOO-AH-MOO-AH*)

1 ounce aged rum
1 ounce oloroso sherry
1$\frac{1}{2}$ ounces pineapple juice
$\frac{1}{2}$ ounce coconut cream
$\frac{1}{4}$ ounce lime juice
Glassware: Chilled collins
Ice: Pebble
Garnish: Mint sprigs and absinthe mist
Method: Measure all ingredients into a cocktail shaker. Whip with $\frac{1}{2}$ ounce pebble ice. Dump into a chilled collins glass and top with pebble ice. Garnish with mint sprigs and mist absinthe over the top.
Attribute: Bryan Tetorakis, The Varnish

PIÑA COLADA

1 ounce aged rum
1 ounce white rum
1$\frac{1}{2}$ ounces pineapple juice
$\frac{1}{2}$ ounce coconut cream
$\frac{1}{4}$ ounce lime juice
Angostura bitters
Glassware: Chilled collins
Ice: Pebble
Garnish: Pineapple chunk and umbrella

Method: Measure all ingredients (except Angostura bitters) into a cocktail shaker. Whip with ¹/₂ ounce pebble ice. Dump into a chilled collins glass and top with pebble ice. Add 4–5 dashes of Angostura bitters on top and garnish with a pineapple chunk and an umbrella.

QUEENS PARK SWIZZLE

2 ounces white rum
1 ounce lime juice
³/₄ ounce simple syrup
1 white sugar cube
8–10 mint leaves
Angostura bitters
Peychaud's bitters
Glassware: Chilled collins
Ice: Pebble
Garnish: Mint sprigs
Method: Gently muddle the white sugar cube, mint, lime juice, and simple syrup together in a cocktail shaker. Add white rum. Dry-shake and dump into a chilled collins glass. Top with pebble ice and swizzle. Generously add 5–6 dashes each of Peychaud's bitters and Angostura bitters over top. Add topping of pebble ice. Garnish with mint sprigs.

Mr. Potato Head
Hyde Park Swizzle: Sub gin for white rum. (Alex Day, Death & Co)

SHERRY COBBLER

2¹/₄ ounces oloroso sherry
³/₄ ounce curaçao
1 orange wedge
1 lemon wedge
Glassware: Chilled double rocks
Ice: Pebble
Garnish: Orange wedge and powdered sugar

Method: Squeeze and drop the orange wedge and lemon wedge and measure all ingredients into a cocktail shaker. Dry-shake, dump into a chilled double rocks glass, and top with pebble ice. Garnish with an orange wedge and powdered sugar.

⊰ CREAMS AND FLIPS (AFTER THE FACT) ⊱

BRANDY ALEXANDER

---

1 1/2 ounces Cognac
1 ounce crème de cacao
3/4 ounce heavy cream
Glassware: Chilled coupe
Garnish: Nutmeg
Method: Measure all ingredients into a cocktail shaker. Add a rock of ice and shake hard. Strain into a chilled coupe. Garnish with grated nutmeg.

Mr. Potato Head
Alexander: Sub gin for Cognac.
Coffee Alexander: Sub Café Varnish (see p. 133) for crème de cacao.

CAFÉ CON LECHE FLIP

---

1 ounce Gosling's Black Seal rum
3/4 ounce Café Varnish (see p. 133)
3/4 ounce simple syrup
1/2 ounce cream
1 egg yolk
Glassware: Chilled sour
Garnish: Nutmeg
Method: Measure all ingredients into a cocktail shaker and dry-shake. Add a rock of ice and shake hard. Strain into a chilled sour glass. Garnish with grated nutmeg.
Attribute: Sam Ross, Milk & Honey

DOMINICANA

1½ ounces aged rum
1½ ounces Café Varnish (see p. 133)
Whipped cream (float)
Glassware: Chilled coupe
Garnish: None
Method: In a chilled mixing glass, stir aged rum and Café Varnish over cracked ice. Strain into a chilled coupe and float whipped cream on top.

NEW YORK FLIP

1 ounce bourbon whiskey
¾ ounce port
½ ounce cream
½ ounce simple syrup
1 egg yolk
Glassware: Chilled sour
Garnish: Nutmeg
Method: Measure all ingredients into a cocktail shaker and dry-shake. Add a rock of ice and shake hard. Strain into a chilled sour glass. Garnish with grated nutmeg.

WHITE RUSSIAN

1½ ounces vodka
1½ ounces Café Varnish (see p. 133)
Whipped cream (float)
Glassware: Chilled coupe
Garnish: None
Method: In a chilled mixing glass, stir vodka and Café Varnish over cracked ice. Strain into a chilled coupe and float whipped cream on top.

⊰ HOT STUFF ⊱

HOT BUTTERED RUM

1^1/$_2$ ounces aged rum
^{3}/$_4$ ounce honey syrup
1 pat of butter
Hot water
Glassware: Insulated glass
Garnish: Nutmeg
Method: Measure the rum and honey syrup directly into an insulated glass. Add the pat of butter and fill glass halfway with hot water. Stir with a barspoon until the butter is emulsified. Top with hot water. Heat with a steam wand if hotter temperature is desired. Grate fresh nutmeg on top.

HOT TODDY

1^1/$_2$ ounces bourbon whiskey
^{3}/$_4$ ounce honey syrup
2 lemon wedges
Hot water
Glassware: Insulated glass
Garnish: Nutmeg and cinnamon
Method: Measure all liquid ingredients directly into an insulated glass. Squeeze both lemon wedges and drop into the glass. Top with hot water. Heat with a steam wand if hotter temperature is desired. Grate whole nutmeg and cinnamon on top.

Mr. Potato Head
Fireside Toddy: Sub ^{3}/$_4$ ounce aged rum and ^{3}/$_4$ ounce Cognac for bourbon whiskey and throw a star anise and two cloves in along with the lemon wedges.

IRISH COFFEE

1^1/$_2$ ounces Irish whiskey
2 brown sugar cubes
Espresso or strong brewed coffee
Hot water
Whipped cream (float)
Glassware: Insulated glass
Garnish: Orange twist
Method: Measure Irish whiskey and place brown sugar cubes directly in an insulated glass. Add shot of espresso and top with hot water, leaving some room for the whipped cream. If using drip coffee, omit the water. Mix with a barspoon to dissolve the sugar cubes. Heat with a steam wand if hotter temperature is desired. Float cream on top. Express oils from an orange twist over the top and discard.

In the end, I love simplicity. It's what I strive for in my bar, as well as in life. But simplicity is not to be confused with thoughtlessness, and complexity is not to be confused with genius: a daiquiri is lime juice, sugar, and rum, and as long as those three simple ingredients are fresh, measured, and shaken with good block ice, there is no reason to search for a more "interesting" offering.

CHAPTER NINE

HOUSE RULES

As I approach The Varnish, I can hear Sari from twenty feet away regaling Big Jay with tales of matrimonial woe, waving her cigarette in the air for emphasis—an actress forever rehearsing her lines. If I owned a bar like my friend's place in Venice, I might take it in stride, actually worry if my staff wasn't smoking like chimneys, waxing unpoetic about last night, and disappearing in the middle of service to play Ping-Pong down the street.* But The Varnish isn't that kind of place. And while I'm pretty relaxed about our behavior in general, I'd prefer to keep smoking and partying and our other vice-like habits hidden from view.

"Sari," I say, surprising her. She automatically moves her hand toward the ashtray. I stop her.

"If you're going to smoke, take a lap around the block, wash your hands, chew a mint sprig, then get back inside."

Big Jay smiles at me in ersatz brotherhood. I notice the neon coupe above his head isn't on. One of his only opening duties.

---

* Surprisingly that bar still exists; not surprisingly, under new ownership.

"Jay, please look up," I say, and without waiting for a response, whisk my way inside.

BROILED MEAT. HOUSE-BAKED FRENCH BREAD. Atomic mustard that burns my eyes. Cole's restaurant stinks like a hunting lodge. Like pipe smoke. Like the ghost of Red Car riders who blew through here in the forties when it was a stop on the trolley. Back then, the place was many things—a home to Mickey Cohen, a watering hole for stockbrokers, and Bukowski's preferred haunt for boilermakers. Today it caters to a mix of old-timers and new, offering a menu and a room design that skews close to the original; it isn't thematic, but takes its cues from the past—textured red velvet wallpaper, leather burgundy booths, hand-painted signs, taxidermy, and globe lights.

As I make my way through the room toward the secret door in the back, trays of French dip roll out of the kitchen like parts on a GM assembly line—brown meat stacked between soft white baguettes, an atomic pickle on the side, au jus sloshing around on servers' trays. A basket of garlic fries teeters on the edge of a tray and flips over onto the floor. Viola barely flinches—she's been here for years, and when she's not waitressing, she's hypnotizing people in an office in the Valley.

"Here let me," I say, but she waves me off.

"I got it," she bends down. It's hard for me to walk away when something goes wrong; I always feel like it's my responsibility, but I have to keep reminding myself that holding everyone's hand in service can be a disservice. That nobody learns if I don't learn to delegate. "I hear they're getting slammed back there," Viola nods behind her. So I stand up and open the door to The Varnish.

"Well, then you have a small penis!" exclaims a five-foot tiara'd bride-to-be.

Now I love a bachelorette party, especially at 2 a.m. with a bottle of bourbon in one hand and a baggie in the other, but at 7:23 p.m. it's like seeing zombies of the apocalypse in the Garden of Eden. Little Miss Betrothed is miffed because The Varnish is unable to accommodate her and her eight friends. Our bar is the size of a three-cent stamp and seats parties of six max, which apparently means that Anthony, our intrepid host, has a small dick.

Anthony smiles at me and mouths "Hopscotch"—Varnish slang for a guest trying to jump the line.

We opened The Varnish to offer people meaningful experiences, to create something that felt like the kind of place someone would want to tell their friends about but also keep to themselves. Our goal was to offer a level of service that inspired guests to sip slowly and savor, leave the troubles of the day at the door or bring them in, and find solace in our service and a glass of something delicious. But creating a space that offers these opportunities to everyone means everyone has to play by the same rules. Here are the ones printed on a sign hanging outside our door: *Welcome, please wait to be seated. Our process takes time; thank you for your patience. We are first come, first served and we don't take reservations. We cannot accommodate parties larger than six people. No vulgar language or loud behavior please. Disrespectful attitudes towards other patrons or our staff will not be tolerated. Cocktail attire is admired, not required.*

Rules, like laws, are predicated on troublesome situations having once occurred, now demanding a remedy. "No shoes, no shirt, no service"?* Chances are someone walked into a joint without shoes or a shirt and was refused service. "If your drink isn't

---

* This sign is seen in many places, sometimes even Florida.

strong enough, ask for a double"* was undoubtedly designed for a customer who complained they couldn't taste their booze. And "During Thursday Turtles Races, never, ever, EVER point at the turtles."† I have no idea what that means but feel certain the rule is prudent.

Our process at The Varnish takes time because we don't speed-pour, but jigger to precise measurements; we spend time with our guests to ensure they're getting a drink that suits their mood; and we build drinks by the round when possible, as opposed to one by one, so that a group of guests can all toast together. We don't take reservations because we're tiny and fill up fast—if we booked our tables, no one would be able to stop by on a whim and get seated. We don't allow people to wait inside, because those seated deserve to enjoy the drinks in front of them instead of the ass of some random, but people push their way in anyway, say, "I'm just looking for someone, I'm not going to stay," but they're looking for no one, and as soon as they see an opening at the bar, sidle up and order a drink. These people make me see red like Carrie at the prom—an extreme reaction I'll admit, but when someone lies straight to my face then stands three feet away as if I can't see them or recall their disregard, I feel murderous.

We don't believe that being under the influence of alcohol means relinquishing one's sense of decorum, which doesn't mean that bars with rowdy atmospheres don't have their place, but The Varnish isn't that kind of saloon. And there is never *ever* a reason for disrespect—of guests or staff. We have a zero-tolerance policy on that rule. As far as cocktail attire goes, Sasha's initial desire at M&H was to create a place where a genteel decorum reigned

---

* The Rail Pub, in Savannah, Georgia.

† Brennan's in Marina del Rey, California.

in sound, taste, and dress. For him, there was nothing "theme-y" about a button-up shirt with suspenders, pressed pants, and shined shoes—just a belief that dressing well contributed to living well. And we all followed suit.

The Varnish's house rules are actually a dialed-back version of M&H's OG rules, which read: *No name-dropping, no star f*cking. No hooting, hollering, shouting, or other loud behavior. No fighting, no play fighting, no talking about fighting. Gentlemen will remove their hats. Hooks are provided. Gentlemen will not introduce themselves to ladies. Ladies, feel free to start a conversation or ask the bartender to introduce you. If a man you don't know speaks to you, please lift your chin slightly and ignore him. Do not linger outside the front door. Do not bring anyone unless you would leave that person alone in your home. You are responsible for the behavior of your guests. Exit the bar briskly and silently. People are trying to sleep across the street. Please make all your travel plans and say all farewells before leaving the bar.*

These rules were created not just to foster a desired atmosphere, but as a requirement for staying open—Milk & Honey was located underneath a residential building, and a certain level of discretion was necessary to keep it all going until dawn.

That said, rules dictating how one should behave in a bar are a relatively new concept. Since the sixth century B.C., Dionysus, the Greco-Roman god of wine and ecstasy, has been the poster child for bacchanalian mêlée. He's depicted in paintings grasping a two-handled goblet, surrounded by satyrs and nymphs in fawn skins, accompanying demonic maenads in forests on the hunt for victims to devour raw. This style of partying formed the basis of my own early behavior, so I get that it's the more common approach. Which means that when customers flout the rules, we do our best to guide them toward a better way before they ruin everyone else's good time. With this in mind, I head back to the front door and the nine bachelorettes in their various forms of disarray and in my most Zen of voices say, "If you split into two parties, I can offer four of you a place at the bar right now and a table as soon as one opens. If the rest don't mind waiting at Cole's, I'll come out and get you the moment something becomes available."

The one with the tiara, the soon-to-be bride (God help her intended) says, "Fuck this place this place sucks," and stumbles past me on her towering heels in a cloud of coconut body oil.

"Hodor," I say to Anthony—slang for "hold the door"—and pour us each a shot. It's gonna be a long night.

MORE VARNISH BAR SLANG—USE IT AT HOME!

→ LUBE 'EM UP: Give 'em a shot.

→ OTF: On the fly; need it fast.

→ GOODNIGHT MOON: A customer's last drink.

→ FOR A COP: Don't fuck it up.

→ SLEEPERS: A table that won't leave.

→ MONEY SHOT: Seriously discount a tab.

→ CHEESEBURGERS: What you say when the health department shows up unexpectedly. Throw away the raw eggs immediately.

→ SNAIQUIRI: A half daiquiri for staff or a VIP.

→ JIBRONI: A half negroni, in the same vein as a snaiquiri.

→ FAMILY MEAL: A meeting at the bar where the staff shoots snaiquiris or shots at the start of service and intermittently throughout the night. They are a way to unite, a reminder that we're all in this together, for better or for worse, for the next seven hours.

→ ASIAN EQUATION: Saturday night bartender team consisting of Daniel Eun and Eugene Shaw.

→ TARBY PARTY: Friday night bartender team for anyone working alongside Devon Tarby.

→ POOH BEAR: Nickname for bartender Chris Ojeda, who was methodical but slow as honey.

→ "EVEN A BLIND HOG'LL FIND AN ACORN EVERY ONCE IN A WHILE": A popular Bostickism from Chris Bostick, former bartender and GM. Usually uttered as a backhanded compliment when you finally got your shit together.

→ "ATREYU!": A server taking a heavy tray with both hands calls out to the bartender to put up a new tray for the next round.

→ SEE SEE: Put on end-of-night playlist. A reference to the song "See See Rider."

→ "PAPA HIELO!": The literal Spanish translation is "daddy ice," which sounds like a Biggie lyric, but by which we mean "Get more ice." Coined for Varnish barback Carlos Lopez-Flores who has worked at the bar since day one, and sober no less. Carlos is the backbone of The Varnish, and the framed article hanging on the wall by the front door from an issue of *Life & Thyme* magazine celebrates him as one of the industry's unsung heroes.

→ DON'T SARI THEM: Do not give resting bitch face.

CHAPTER TEN

YOUR LOCAL

A local bar is a port in a storm. A place where *everyone knows your name,* and not just your name but where you work, what your relationship status is, and your preferred drink at various points throughout the night. "Local bar" is often synonymous with "dive bar," but this isn't always the case. The principal characteristic of any local—be it "fancy" or "divey"—is that it loves you unconditionally, so when you stand on top of the bar and proclaim yourself a golden god in the middle of Friday night service, promptly get 86'd,* then spend the following day dressing up in disguise and returning to the scene of the crime,† your local will slide a PBR across the bar and (begrudgingly) welcome you home.

Friends you make in your local become friends for life. Falling in love in your local is encouraged, but falling in love and then breaking up and then fighting over who "gets the bar" is verboten. You might marry someone you meet in your local, and that

---

* Barred forever.

† True story.

guarantees your photos will be stuck on the wall behind the register. When you lose your job, your local will run a tab for you until you get back on your feet, and after hours it will let you hang out while it counts its cash and washes its dishes, dims the lights and puts you in a cab home.

But when someone asks what it is about your local that makes you love it so, it usually can't be explained. No single thing you can put your finger on exactly; it's just an overall *feeling*.

In 2011, I helped my friends Craig and Annie Stoll open their Roman restaurant, Locanda, in San Francisco. One night soon after opening, Craig and I were sitting half-dead on the couch, half-drunk beers in danger of spilling their remains, when he opened his eyes and said, as if midconversation, "The dining experience is made up of hundreds of tiny details that, with the right touch, come together to transcend the sum of their parts. But you need that right touch, or magic . . . that third rail of love, to make it jell."

That third rail of love is the indefinable element that defines a local—the invisible woo-woo that pulls everything together to make a place feel like yours. But in the same way you can't set out to fall in love, you can't set out to become a local. There are factors you can never anticipate that make someone crush so hard on your bar they tattoo its logo on their arm.[*]

What you *can* start with are "the three pillars." Concept: Is yours realized? Space: Will your concept work in the space? And location: Is it a popular neighborhood, or will you be pioneering an up-and-coming area?

Out of the gate, The Varnish had two of the pillars locked in—a verified concept from M&H/Little Branch and a space that fell in line. Our challenge was location. Los Angeles, in particular

---

[*] See Afterword.

downtown, is experiencing one of the worst homeless crises in the United States. In 2009, the neighborhood was not a destination hot spot. It was, and still is, seedy and rough. When you open a bar or a restaurant or a store that people have to travel to get to, everything they see and interact with along their way informs how they'll feel upon arrival. Weathering the conflicting conditions of downtown's mash-up of rich and poor, new and old, fucked up and fancy, was for sure going to affect how people felt when they reached The Varnish's front door. We didn't want to ignore that; we wanted to celebrate the history and authenticity of the neighborhood— attributes that can be difficult to find amid L.A.'s more prevailing artifice. Downtown and its denizens were here long before us—it was our duty to add something genuine and thoughtful to the mix.

In tucking The Varnish into the back of hundred-year-old Cole's, the goal was to create a natural segue from the dining area into the cocktail bar, following the Victorian design influences and reworking only what was necessary to keep things operating. It is an honor to be gifted a space that has stories hidden inside the walls. Literally. There's a peephole next to the bathrooms that reveals a stairway that once connected Cole's and the train platform above. A trio of old suitcases and a fedora lie across one of the steps, as if the owner was just out of frame, buying a ticket. I don't know if someone placed them there for theatrics or if they're an actual relic from another time. I like to think the latter.

We made The Varnish's front door heavy, wooden, and nicked all to hell, with a bulbous brass doorknob that doesn't turn. It's meant to feel well-worn, old, and magical.

The bar's rectangular space is laid out with a standing bar directly to the left of the front door and booths arranged in a horseshoe around the room. There are three café tables by a pony wall, aka half wall, separating the seated area from the standing. These spaces are discreetly separate on purpose, because have you ever sat in a chair while someone stood next to that chair talking to someone else? Very awkward.

The overhead warehouse lights are encased in steampunk cages that lower the ceiling to create a more intimate feeling than something lofted. Our comfort zone is New York—cramped, low spaces with tables so close, guests have no choice but to socialize.

The sconces at each booth have seven-watt bulbs that bathe those seated in a diffused, soft-focus Hollywood glow. But it's the concrete column in the center of the room that I truly love—a wraparound marquee light that stands like the monolith in *2001: A Space Odyssey*,* with a killer custom light fixture made out of an inverted rain gutter filled with marquee bulbs. Ricki did good on this one.

Behind the bar, rope lighting adds a shimmer to the bottles on the shelves, and on the underside of the work stations are 2,700-kelvin warm white work lights. The last thing I want is to have a beautiful glow going across the room and then blast the bar with cold, dead light. The only "bright" lights in the room are the two MR16 halogens over each work station that cast a spotlight on the bartender's hands in action, like a magic show, contributing to the overall sense that this is theater.

Even elements that seem ancillary play a role, like the hooks. Hooks. Are. So. Important. There are hooks under the bar and at the tables for people to hang their coats, hats, bags, and scarves on. Such a tiny thing, but have you ever gone someplace and rooted around for a hook, swiped your hand back and forth under the table or along the wall, only to find nothing? Where do you put

---

* Stanley Kubrick's 1968 sci-fi space thriller. If you don't know this movie, stop reading this book, eat some acid, and watch before continuing.

your coat? On the back of your chair where it's liable to slip off or, if it's long, get stepped on? You could drape it across the back of the booth if you're in a booth, but then it's also hanging into the booth behind you. And what about your hat? That fucker's getting crushed if a bunch of you are squeezed in. And bags are annoying to hold on your lap when you're meant to be relaxing and letting everything drift away except for the company you've chosen to keep and the drinks you've ordered to sup.

Establishments that don't have hooks have not thought through their guests' experience. And it's super cool to make them look and feel nice, too, because why not? Ours are oil-rubbed bronze octopus hooks with legs that swing upward like their namesake cephalopod. They cost $3.05 each at Home Depot and have that weathered, lived-in feel from another time, and, while no one sees them, they're lovely to touch. Textures are as much a part of the guest experience as anything. In a dark bar, under the influence, people feel their way through the space with their hands a lot. The varnished walls, leather rail, and brass doorknob are all sensual points of contact.

Let's discuss POS (point-of-sale) systems. The Varnish has an old 1960s cash register that came with the bar and sits dead center, to ring in cash. It's an interactive museum piece we find charming; however, we need digital POS terminals for credit cards and these must be hidden from view. First off, they're visually offensive— who goes out to be accosted by a computer screen from across the bar? And second, the transaction part of a guest's experience doesn't need to be on display. When a guest wants to pay, it should be done quickly and out of sight. Like a drug deal.

Then there are the tchotchkes—trinkets and collectibles we've scattered around the room that add a lived-in, personal touch: antique punch bowls, teacups hung on the wall, a 1940s black rotary phone, framed family photos. There's one of my mother at age fourteen in black-and-white, hanging on the concrete column next to the piano. She's posed in front of a ribbon microphone inside a sound booth, where in 1956, she recorded "Je Dis Que Rien Ne M'Épouvante"*—the innocent village girl's lament from Georges Bizet's *Carmen*. My mother doesn't often talk about her life as an opera prodigy in Morocco. She graduated first in competitions at the Conservatoire de Musique in Casablanca, and hosted a weekly televised talent show wearing a white dress with light yellow stripes her mother sewed. Family matters and Morocco's political post-independence volatility shifted her priorities, and her dreams of musical stardom were replaced by the more practical career choice of joining forces with her mother and brother Michel to open a restaurant in El Port de la Selva, Catalonia, Spain, called La Estrella del Mar.† But opera was her first love, and my earliest aural memories are of Mozart, Léo Delibes, and Bizet; the voices of Mado Robin, Renata Scotto, and Maria Callas. In the 1980s in Mamaroneck, New York, my mother would press her ear up against our Pioneer SX-1980 and sing along with the spinning records, her coloratura soprano brighter and more beautiful than anything in the room.

The music we play at The Varnish is an extension of this. In the same way the furniture and lighting, penny tiles and worn front door were curated to hit a mood, the music has been designed to evoke a feeling of nostalgia, love, and warmth, featuring a wide swath of jazz styles, bossa nova, lounge, Motown, doo-wop, and French and Italian pop spanning from 1920 to 1960. On particularly busy nights we might spike the vibe with a few outliers, but overall we keep the sound consistent, unless it's a holiday or our

---

* "I say that nothing frightens me."

† The star of the sea.

anniversary and then all bets are off. Our Spray-tanniversary boasted Long Island Iced Teas, Adios Motherfuckers,* Jäger shots, and top forty. Our seven-year formal served up Beefeater martinis and Champagne with our house musicians acting as a live crooner karaoke band. One Halloween we threw a *Young Frankenstein* party featuring cocktails based on the movie, and the ragtime-Ziegfeld-Broadway sounds of Irving Berlin.

Some bar owners believe one should "play to the room"—meaning the music should be tweaked to appeal to those present. I would argue that a bar or restaurant or shop or any space where guests are invited to linger should play to the *design*. Maybe the proclivity for giving guests sonic control is left over from the days of jukeboxes, when anyone could sally up to the Wurlitzer and pick out all the songs a quarter could play, but that is, somewhat sadly, no longer the jam. Except for places that *do* still have jukeboxes, in which case the democratic choosing of songs is an agreed-upon conceit.

But—and there's always a *but*—it's pretty cool when a bar hands the musical reins over to its employees. There are few better ways to imbue your staff with a sense of ownership than letting them choose the tunes during their shift. It's also a cool way for them to share a piece of themselves, to let their neo-soul, classic-rock, Mexican-rap, thrasher-skater, punk-pop freak flags fly.

"But that seems contrary to everything you've just stated," you might argue.

Except not really. The idea is that it doesn't matter *how* you curate your music; it just matters that you do. It matters that it's considered in the same way every other aspect of the bar has been considered. Which means that a human is at the helm. Which

---

* A Long Island Iced Tea with blue curaçao and 7-Up.

means that never, ever, ever, ever, *ever* should you let an algorithm choose your music. The rampant predilection for letting Spotify and Pandora pick your songs is a true heartbreaker. Play whatever you want in your bar/restaurant/coffee shop—it's your prerogative! But relinquishing control to an algorithm? Do you let an algorithm select the liquor for the backbar? Do you let an algorithm determine which bakery you buy your bread from? Do you let an algorithm choose the beans for your brew? I'm going to go out on a limb here and say no. No you do not. And please believe me when I tell you I love AI. I tell my computer to shut off my lights, raise my blinds, and read me to sleep at night. But when it comes to curating a vibe, that is a job for real intelligence.

And if opera on the turntable in Mamaroneck is the root of my affection for collecting records, my love of seeing live music was firmly established the summer I spent at Espace du Possible.[*]

I was twelve years old when my parents, hoping to stimulate my dyslexic brain with new sights and sounds, sent me to Paris to stay with my tonton[†]—the aforementioned Michel. After picking me up outside the Charles de Gaulle airport in his vanilla-colored Citroën 2CV, he drove us to a "naturist resort" on the Atlantic near Royan, in Meschers, where, unbeknownst to my mother, he had been living for months.

During the day, (a clothed) Michel hung out with me and the rest of the (clothed) kids, arranging Olympic-style games complete with medals and ribbons. When he wasn't around, he was off doing whatever it was nude adults did, which we under no circumstances wanted to know about. Left to our own devices, we built forts, threw pots, and rode tandem bikes, while at night we

---

[*] The place of possibility.

[†] Uncle.

huddled around campfires eating Petit-Beurre cookies and smoking cigarettes, looking considerably less cool than we imagined.

Some nights, our presence was required at the talent show.

The adult talent show.

If ever there was an experience inessential to a young person's development, it's watching a camp full of naked adults playing cover songs. Bob Dylan. The Rolling Stones. Rod Stewart. Us kids would show up with our families, and as soon as we could we'd flee, find each other in the crowd, and escape under the cloak of night.

But one night it was just me, sitting alone, enduring the nude musical revue without respite, when a voice in the dark whispered, "Viens ici."* I looked up to see my friend Manon standing in front of me, holding out her hand. I placed mine in hers and rose. This in itself wasn't particularly notable—the French are a demonstrative lot, kissing, hugging, playing the tuba naked—so to find myself running hand in hand with a girl across camp and then jumping into a hot tub wasn't *totally* strange.

Or so I kept telling myself.

As the only French American in camp, I had to be careful not to do anything gauche.

As we submerged ourselves in the water, a French woman started singing "Angie," as if in accompaniment.

Angie, you're beautiful

Water rose in undulating rivulets up to our chins and then receded as the wake our little bodies created settled.

But ain't it time we say goodbye

All it took was a slight turn of my head to join our lips, and just like that we were kissing—*roulant la langue!*†

* "Come here."

† With tongue.

If having your first French kiss be with a French girl in a hot tub in France while a naked French woman sings the Rolling Stones isn't an indelible blueprint for the rest of your life, I don't know what is.

You know how people ask "Beatles or Stones?"

Stones all the way.

Which brings me somewhat tangentially to live music at The Varnish. Nothing short of my own little slice of heaven.

Jamie Elman on our temperamental P.S. Wick, Johnny Sneed on drums, and Marc Gasway on upright bass are our core crew, with other notable musicians dropping in on the reg to play alongside. When it includes a brass section, The Varnish becomes the Prohibition-era jazz club of my fever dreams, the air vibrating with strings, the snare snapping, the piano grinding, servers, bartenders, hosts, and guests, working, moving, talking, and laughing in concert.

For all the artistic expressions that move us—sculpture, movies, painting, design—music is one of the rare things we can actually *watch being made*, and watching an artist in the throes of creation is outrageously special. It's not like you can just walk into Damien Hirst's studio and hang out while he adorns human skulls with diamonds or show up on set and watch Robert Downey Jr.'s process. You can't pop over to Tom Ford's atelier to study how he designs suits. But you can buy a ticket to a concert and watch a musician you love make their art in real time. Deadheads dancing around in ankle bells and caftans, holding hands and singing about morning dew, punk rock fans slam-dancing and stage-diving into each other's arms, classic rock audiences singing every lyric to every song—the sounds and expressions may change, but the catharsis is the same. The brilliant writer Hanif Abdurraqib explains it beautifully: "What I'm mostly invested in is the ecosystem or the geography of a room beyond the stage, where everyone's emotional impulses are kind of linked by the thing that brought them there tonight—that idea that audiences could see something potentially magical and impossible, and what that does to a room, what that does to the people in a room, and how people share that and hold that space with each other, for better or for worse. To me, that is more interesting. The kind of tactile human nature of concerts. And I think that's really valuable."

Live music at The Varnish is meant to be an accompaniment to the night, not the sole focus, but an important addition nonetheless. An addition that changes the vibe of the room and the people in it, linking them whether they know it or not. When we have live music at The Varnish, the place feels complete.

ON AUGUST 21, 2015, I flew to Australia to visit my friend Michael Madrusan, to consult on his new ice company and attend the opening of his bar, Heartbreaker. Michael was partners with Sasha on the incomparable Everleigh in Fitzroy, and the two of us knew each other from our days together at Little Branch. Fifteen hours after takeoff, I landed in Melbourne to three voicemail messages on my phone telling me that Sasha had died. I stood in the middle of the airport in shock, my carry-on heavy against my shoulder. To my right were signs pointing me toward baggage claim; to my left were ticket counters where I could purchase a flight back home.

My jet lag made it all feel very surreal. *Maybe I'm dreaming,* I thought. *Maybe this is a nightmare I'll wipe from the corner of my eyes once I catch a breath of fresh air.* Except no matter how long I stood there staring at my phone, the messages didn't disappear, and I had to make a choice—should I stay or should I go?

I stayed.

That night, Michael and I got LOVE KILLS tattoos. The following night, we bartended the grand opening of Heartbreaker. And in the morning I rented a BMW F800GS—a long-distance adventure touring bike—and rode the Great Ocean Road, tears flowing all the way to the Twelve Apostles.*

"Left, left, stay left," I chanted as a reminder to stay on the correct side of the road. When I looked out on the grandeur of the Australian coastline, my mind wandered back to three days earlier, when I texted Sasha about a cocktail spec. I asked him how married life was and he told me it was tough. A few months before he got married, Sasha asked if I would buy him out of his shares of The Varnish so he could cover his wedding expenses—a ring, a suit, a honeymoon on the Orient Express. As usual, he was broke. He was also in a rush because his mother, Anita, was ill, and he wanted her to see him wed before she passed.

Georgette and Sasha had a May wedding. Everyone who was a part of the M&H family was there. Three weeks later his mother died, and soon after that Sasha was found dead from a heart attack. He was forty-two years old.

I rode thinking how the man who created careers for me at The Varnish, for Joe Schwartz at Little Branch, Richard Boccato at Dutch Kills, Michael Madrusan at The Everleigh, Sam Ross and Micky McIlroy at Attaboy, and hundreds of others had been struggling all along. Sasha helped pave the way for us to build bars that shared a connection beyond specs, and resonated in our relationships with one another. From New York to L.A. to Australia, a line bound us together that began with him.

At some point—I can't tell you when, only that my body and

---

* A collection of limestone stacks on the coast of Port Campbell National Park, in Victoria. The Apostles were formed by erosion from the harsh weather conditions on the Southern Ocean. There are eight that remain, but the name has stayed the same.

brain needed to stop running—I spotted a small green sign that read HISTORIC BOGGY CREEK PUB—3 KILOMETERS.

I made my first right turn on a trip that had pointed only southwest, rode through a chain link of dairy farms, and found myself in the small town of Curdievale, population 124, and the Boggy Creek Pub—the most pastoral local I'd ever seen.

I dismounted and pushed through the heavy worn wood door into my record-scratch moment, only to have the silence broken by four young kids hollering at Australian Rules football. They must have been eighteen or nineteen years old (drinking age is lower in Oz), and when they saw me, bought a round of pots* for us all. The sound of the commentator over the television battled with the jukebox's top-forty songs. Wood-paneled walls and a yellow-tiled ceiling; the shelves above the backbar were filled with statues of local ducks and birds. The pub was family-run, and I met the older son over my first Carlton Draught.† By my fourth, the mum showed me to a room in the back where I could stay the night, and when I returned to the bar before retiring, the dad had taken over tending.

Dairy farmers, fishermen, seventies hits, and more pots. I ate the Coulotte steak frites, and at midnight, cozied up in the chilly country back room and finally laid my head.

The world is a big, bad place. It's also a tiny, miraculous place. Wonderful things happen you could never imagine, and terrible things happen just the same. Perhaps the most life-affirming moments are when you're at your darkest and someone opens their arms to you. Someone you don't know. Someone who just looks at you and, even though they don't know the why, know you need

---

* Different regions in Australia have different names for their drinking vessels, but in Victoria and South Australia, the large (450 milliliters/15 ounces) is a "pint" and the small (285 milliliters/10 ounces) is a "pot."

† A 4.6 percent ABV beer made in Australia by Carlton & United Breweries.

comfort. It can happen in your neighborhood, and it can happen on the other side of the world in a tiny, dark tavern filled with people you've never met. People who've created a space for locals and travelers alike to be gifted with the intangible third rail of love.

My directionless search for soul comfort over losing Sasha brought me to the Boggy Creek Pub, a foreign place in a foreign land, at one of the lowest moments of my life. But it was there that I was reminded of what our lives together—mine and Sasha's—were about. We wanted to be that port in a storm. We wanted to comfort and mend. Celebrate and serve. We wanted to make a place where people felt at home.

I rose early to the frost and hit the road.

CHAPTER ELEVEN

TOP TEN REASONS NOT TO DATE A BARTENDER

People who are attracted to working in bars are fringe people. People who can't sit still. People who can't wake up early. People who hate to use their inside voice. Extroverts, exhibitionists, and entertainers—the qualities that make a good bartender are not the qualities that make a good boyfriend or girlfriend.

While a bartender's life is the life I have chosen, one I truly believe has honor and merit, I know that being in a relationship with a bartender can be a challenge. So for anyone who has ever considered, is currently considering, or finds their future self contemplating dating a bartender, here is a handy reference guide of things to keep in mind.

1   YOU NEVER SEE EACH OTHER. Normal people work nine to five, Monday through Friday. Bartenders work 4 p.m. to 2 a.m. some weeknights, and most weekends. Your window for delightful farmers' market adventures, sex, and general hanging out lies somewhere between the inarguably unromantic hours of three and eight in the morning.

2 THEY KNOW EVERYONE. When you go out to dinner you'll feel like a celebrity—all the free drinks, gifts from the kitchen, diners turning their heads wondering who you are. Half your meal will be comped, i.e., free, and service will be off the chain! But half the duration of your meal your bartender will be gone, disappeared into the kitchen with the chef, getting lost in discussions about custom stainless shelves, Cryovac bags, and butcher paper. When your bartender returns, the sous-chef will join you at your table, sitting down to make sure you're enjoying your meal. They will stay a long time. After dinner, just when you're ready to leave, your bartender will get pulled back behind the bar to check out the ice prep station, the glass chillers, and the sliding ladders, talk about the drinks you had at dinner, go over some specs and drink an amaro. One day, you'll ask if the two of you will ever enjoy a quiet, undisturbed evening out alone.*

3 THEY DRINK TOO MUCH. Bartenders drink like chefs eat. *A lot.* Between the family meals, the shift drinks, and the shots bought by customers as a form of liquid love, a bartender's night is pretty saturated. Even if your bartender doesn't start out as a big drinker, they're either going to become a big drinker or drink too much for a nondrinker, because it's nearly impossible to keep saying no. And the more your bartender says yes, the more drinks they're offered, because people want their bartender out of control like Lindsey Lohan in a car chase down the PCH. The wilder the better! It's all part of the circus—an impressive trick to juggle drinking and laughing and pouring and lining up the next round at the same time. The adoration your bartender feels these nights will become a drug. Speaking of which . . .

---

* You won't.

4   THEY DO TOO MANY DRUGS. People go out of their way to share drugs with your bartender. The drawers in their office overflow with baggies of mind-altering substances. The pockets of their jeans have fine dustings of white powder embedded in their seams. One Christmas, on the way to meet your parents, you will watch your bartender walk through a TSA full-body scanner at the same moment they realize there's an eight ball in their back pocket. After that, you will conduct a full military-style inspection of all carry-ons before reaching the gate.

5   ALL THEY DO IS TALK ABOUT WORK (PART 1). If you're also a bartender, it can be great dating someone sympathetic to the long hours you spend on your feet, the way your work clothes wear at an accelerated speed, and how truly golden silence becomes when you're off the clock. Pretty soon, though, your brain will melt, your universe shrinking to the small-world vagaries of staff personalities, brand ambassadors, and preferred scheduling. You won't see movies. You won't read books. And until your friends drag you to the Griffith Observatory for an intervention, you'll forget there's an entire universe out there waiting to be discovered.

6   ALL THEY DO IS TALK ABOUT WORK (PART 2). If you're not in the industry, when you ask your bartender how their night was and they respond with tales of guests finger-banging at table 5, co-workers blowing lines in the bathroom, and regulars puking in a booth, you'll be hard-pressed to respond supportively. You're no prude, but if office hours entailed watching clients make out with each other, walking in on co-workers snorting rails in the bathroom, and cleaning up after your boss puked in the middle of an important meeting, you'd be looking for another job. So, while you appreciate the passion your bartender has for serving others, you'd appreciate it much more in the form of, say, a doctor or a lawyer.

7   THEY WANT TO BE TAKEN CARE OF. All hospitality jobs are performative, but a select few are done in front of an audience without respite. Guests can see every move a bartender makes. Hear every word they're saying. It's like being onstage for hours on end without a script, and you must improvise flawlessly and serve efficiently. When someone spends their nights in this constant state of *ON*, it follows that when they're *OFF*, they want to shut down. Be taken care of. But relationships are a two-way street, and while a bartender feels like they're the one always giving, they aren't always giving to their significant other (see: you). And when you try to explain to your bartender that no, they don't have to do much, but yes, you want their love and attention, well by the time this conversation happens . . .

8   ALL THEY WANT IS TO BE ALONE. Your bartender is surrounded by people all the time, and not just people but the worst kind of people—drunk people. Mostly what he or she longs for when not working is a daybed, a bag of 50:50 sativa/indica hydroponic Grease Monkey with purple trichomes and complex terpenes, and a Netflix show with infinite seasons. But after waiting patiently for your bartender to have a night off, you want to do something fun, go someplace together! But that plan doesn't sound so compelling to your bartender, who by now is so stoned they couldn't possibly go out in public.

9   CHEATING. Now I'm not saying this is a fait accompli, but when someone spends long hours, late at night, surrounded by free booze, great tunes, mood lighting, and innuendo-laden conversations, the deck is definitely stacked in favor of indiscretion. Industry events in exotic locales, rooftop launch parties, and the Spirited Awards* multiply the po-

---

* The bar industry's Oscars, given out at Tales of the Cocktail.

tential of sexual misadventures by a thousand because *everyone's* a bartender, and *all the booze* is top-shelf, and the self-congratulatory mood is flagrant and reciprocal. And cheating isn't always defined as straight-up sex; there are flirty conversations with people who aren't girlfriends or boyfriends when dropping drinks off at their table, casual brush-ups against co-workers in the tight space behind the bar, and longer-than-necessary hugs goodnight with a guest who has stayed after last call.

But it isn't all doom and gloom dating a bartender, because a lot of the bullshit that makes a bartender unappealing as a lover are the very things that make them . . .

10   SIMPLY IRRESISTIBLE! Your bartender is most likely charming and funny and fun. You love them! You want to spend time with them! Well absence makes the heart grow fonder, so if your bartender isn't around a lot, the times you *do* see each other are that much more exciting. Like when you finally take that vacation to Thailand, which is being paid for by a major liquor brand in return for your bartender slinging drinks at a full-moon party on Tonsai Beach—well fuck it if the novelty of being alone together doesn't feel positively illicit. Sex in the ocean, sex on the beach (the act, not the drink), sex in your bungalow immediately following a Champagne-and-oyster lunch . . . When you return from vacation, your friends will ask why you don't have a tan and the two of you will just smile and decline to respond.

In the end, this is really nine reasons not to date a bartender and one list of ways to turn the negatives into positives.

And the free food and drinks when you go out don't stop when you go out *without* your bartender. Those same places will set you and your friends up when you visit and take care of you when you dine out alone. You'll feel like a star even without your bartender/lover, which you should anyway, but free stuff doesn't hurt.

As far as drinking too much and doing too many drugs go, after a while your bartender will stop. Or they won't. And then you should leave them. But most do. Most bartenders grow up and get serious about their craft and segue from bartending to other roles in the industry. They may start consulting for restaurants and bars, become a brand ambassador. They may even consider opening their own bar, which, well, that's a whole other top ten list . . .

Best of luck to you!

CHAPTER TWELVE

ALL ARE WELCOME

When you grow up dyslexic, you hear the word "special" a lot. "Eric's in Special Ed class because he's dumb" was a common refrain. I was actually in Resource Room—a one-on-one teacher/student class for dyslexics that was one door over from Special Ed. From kindergarten through eighth grade, I weathered a lot of shitty social interactions determined to squash my ego, never understanding why kids were so mean.

In high school, things got better when I was forced to join PACE* (thanks Mom), and sophomore year I starred in my first play, *The Madwoman of Chaillot*, as Pierre the young lover who had to kiss the young maid in the fourth act. She was played by Nicole Lipkin, who was a junior with a big football-playing boyfriend, which did wonders for my social standing with everyone, including the jocks, who thought kissing another man's girl was pretty cool.

By senior year I was practically popular, enjoying my role as the on-air anchor for *MHS Information*—our homeroom school

---

* Performing Arts Curriculum Experience for theater, dance, and media arts.

news—which was broadcast daily on the class televisions and on our town's local channel 34.

Despite my John Hughes ascendance, I remained a weirdo, and my friends were the nerds, goths, and Euro kids who talked funny.

At The Varnish (and in my personal life), I have been known to harbor great love for the outcasts, charmed by, and attracted to, the grown-up versions of my childhood friends. I definitely gravitate toward people with dark edges and heavy histories, the ones with tales to tell and the scars and tattoos to prove it. I welcome these individuals into my life and my bar, the same as I do those who are quiet and calm and have good table manners. I'm not saying I'm looking for derelicts to fuck my shit up, but I'm open to anyone, in any shape or form, hanging out until they prove themselves no good.

Sasha always said everyone was welcome at the bar once—that no one deserved to be profiled, and after being given a chance, it was up to them to behave in a way appropriate to their environment. To me, that's hospitality—offering everyone the same level of service regardless of what they appear to be bringing in with them. All of us are guilty of making snap judgments based on how someone looks or sounds, or what our mood is that day, and it's an uncool thing to do in life but an *unacceptable* thing to do in hospitality. Staying open to whoever walks through our door is a form of empathy, a baseline quality that informs every interaction we have.

My good friend Adam Weisblatt, who owns Last Word Hospitality once said, "Service is the technical framework in which hospitality can exist. Service doesn't require empathy, but hospitality demands it." In life you can be empathetic by listening to others, but empathy in hospitality means being proactive and not *reactive*, anticipating a guest's needs instead of reacting to their orders (but, of course, also reacting to their orders). Something as simple as changing out a wet napkin for a dry one, writing down a list of your favorite spots for a tourist, or overhearing it's someone's birthday and presenting their next drink with a flame. It means listening to what a guest is saying and extrapolating what they're *really* saying. Because in a bar (in life in general, but more so in a bar where alcohol has a tendency to fuel exaggeration), the stories people tell are often masks for how they're really feeling. Someone going on and on about how much money they make was probably just fired. Someone super quiet who stays at the bar awhile is most likely desperate to meet someone, and someone boasting about a mad sexual conquest in a super-fast patter while manically twirling a straw is insecure but dying to connect.

The goal at The Varnish—and in any bar for that matter—is to connect by serving others. Of course, every night is different, so the only way to truly deliver on this is by listening to what guests are saying verbally but also physically: Are they taking off layers because they're too hot? Are they leaning into their friend because they can't be heard over the music? Have they pushed away their drink because they don't like it and hope someone will notice but don't want to cause a scene? Each visual cue offers an insight into that person's needs and what you can do to be of service. This attention to detail is how you develop "offhand excellence"—one of my favorite of Sasha's terms.

"In the end, what's most meaningful is creating positive, uplifting outcomes for human experiences and human relationships," Danny Meyer says in his book *Setting the Table*. "Business, like life, is all about how you make people feel. It's that simple, and it's that hard."

It's that simple because what could be simpler than making someone feel good? It's that hard because everyone feels good in different ways. And while every person who walks into a bar is looking for connection, we don't always get what kind of connection they're looking for right. The key is giving a shit. It's easy enough to rejigger an approach if the initial one goes awry—when I think a guest wants to joke around and they turn out to *not* want to joke around, I leave them alone. If I assume a couple wants to be left alone but notice they're looking around the room—maybe they're on an awkward blind date, breaking up, or involved in something much more complicated—I'm going to do my best to help assuage their pain through thoughtful and engaging service.

What drives me *truly crazy* is when a bartender ignores a single person sitting alone. While not all solo drinkers go out looking for their version of *The Hangover*—sometimes they just want to post up with a cold one and a book and be left alone—most people alone in bars would like to meet other people. And in a bar, the bartender is the best one to facilitate this.

The summer of '97, I was backpacking around Europe, when my train made a stopover in Hamburg. It was 11:30 a.m., and the train wasn't leaving again until 11 p.m. I'd never been to Germany before, so I did the best thing I could think of and posted up at the Irish Pub in the Fleetenkieker for a stein of beer. It was right next to the Hamburg Hauptbahnhof railway station, and the narrow, busy bar felt like a good place to experience some local life. The bartender was poetry in motion—easily grabbing orders from the length of the bar, managing several conversations, calling out to guests seated at tables toward the back . . . I snagged the last stool on the short return edge of the bar, next to the wall.

"Hello American!" the bartender cried.

"Is it that obvious?" I asked, more amazed than ashamed.

"But you could have been French, too," he quipped.

"Actually, I'm half French," I told him. "My mother's side. On my way to Paris and I've got ten hours to kill."

"Well, I'm only here for five. What shall we start with?"

Several beers later, he introduced me to his friend Stefan, who was dressed in fatigues, on leave from the German army. I asked what I could possibly see in the city in just a few hours.

"Meine freundin holt mich im tegenbogenauto ab und nimmt uns mit auf eine tour," he said.

Translation: "My girlfriend is coming to pick me up in the rainbow car and will take us on a tour."

His girlfriend's rainbow car turned out to be a Volkswagen she had won in a contest, with different-colored panels from the Volkswagen lineup. This is how I found myself seeing Hamburg through the window of a kaleidoscopic car driven by a giant, beautiful blond woman in the short hours I had before my night train. We walked through the Speicherstadt, aka City of Warehouses, and along the canals of the Elbe River, taking in the gabled facades of the redbrick buildings. We had a picnic in Planten un Blomen, which is Hamburg's version of Central Park, and spent the early evening hours in St. Pauli with its graffiti walls, neon signs, and porn shops. Our last few hours were spent in a strip club before I returned to the station to catch my train.

Thanks to the Irish Pub in the Fleetenkieker's hospitable bartender, whose name is lost to history, I got the best tour of Hamburg a tourist could hope for.

WE'VE GOT A LOT OF incredible bartenders stateside too, tons of men and women who can make anyone feel right at home, but I have to say that my friend Forrest might be the most *masterful*. This guy can find something complimentary to say about everyone. Literally. A guest can be the most obnoxious, worst-tipping, most ridiculously dressed person in the bar and he'll tilt his head and smile and say, "She has great posture."

As a bartender, Forrest is not particularly fast. He isn't the most knowledgeable about what he's pouring, and his attention is easily diverted—it is not unusual for him to disappear with a girl or two or three when he's supposed to be working. Forrest doesn't have a schtick—no witty banter, super flair, or Soup Nazi demeanor that keeps people coming back for more. When Forrest is behind the bar, he's slow and steady; always happy and smiling and charming in a truly understated way. He makes every single person feel beautiful. He treats everyone with the same respect and courteousness, never draws attention to himself, and remembers all sorts of minutiae about people's lives. And he never makes assumptions. Unlike what happened one night at The Varnish when I wasn't officially working, but found myself doing a sweep around the room. A gentleman at table 8 got up and approached the bar. I intercepted him. He told me that he and his friend had been waiting for their drinks a long time and wanted to know where they were. I immediately pegged him as a difficult person, impatient, someone looking to cause a scene. I said, "Good drinks take time," and walked off.

A few minutes later, Kimmy came over with two fresh drinks on a tray and asked where they went. They'd left, because instead of answering their question with a thoughtful "Our drinks do take a little longer, but let me see where they are," I gave a short, pompous answer. Hell, I wasn't even working, I could have walked the guy over to the bar, shown him the service bartender in action, and explained our process, but instead, I lost the opportunity to offer a great experience. Wherever that guy went next, he no doubt badmouthed The Varnish, mocked the uptight owner, rolled his eyes at how long the drinks took, craft cocktails suck, fuck this bullshit I just want a beer and a shot.

And he would be totally justified.

I tell this story in trainings because it's a situation I never got to correct.

⊶⊸◉⊷⊸

SERVING WITH A SENSE OF urgency and empathy is a must in hospitality. An apathetic server or bartender is truly uncool. Making your guest feel like you don't give a shit makes them feel like shit. There's a very real energy transfer that happens between people when they interact. You smile; I smile. You frown; I frown. The latter is not the kind of energy people leave home for. You've probably stood at a bar too long before getting served. Waited way past your reservation to sit down. Ordered a meal that didn't arrive, even as everyone around you who sat down *after you* was happily eating. In all these instances, you probably wanted to ask what the fuck was taking so long but your server/host/bartender just kept avoiding you, and while you were initially annoyed about the wait, grew even more perturbed that you'd been made to feel invisible. Chances are, if someone had just apologized and explained what the holdup was, you would have been all right waiting a bit longer.

As in much of life, communication in hospitality is key. We train hard to make things run smoothly, but shit goes wrong, and owning it and then finding a way to make it go right is where we can make a difference. Being proactive instead of reactive can almost always turn a bad situation good.

WHEN I WAS TEN, we moved from Manhattan to Mamaroneck, but we'd take the train in every month or so to see shows or visit friends or eat at our favorite Cantonese restaurant, Wong Kee, on Mott Street and Hester. Wong Kee was old-school Chinatown—warped linoleum floors, pink plastic tabletops, fluorescent lighting, and a menu with hundreds of dishes all priced astonishingly low. The owner loved my mother. Maybe it had something to do with her being French, and as such, also a "guizi," and that, plus our loyalty as customers endeared us to them. Also, Chinese restaurants are a sort of haven for Jews. Rabbi Joshua Plaut, author of *A Kosher Christmas: Tis the Season to Be Jewish*, says the love affair between Jewish and Chinese people began at the end of the nineteenth century on the Lower East Side, where Jewish and Chinese immigrants lived in close proximity. "The Chinese restaurant was a safe haven for American Jews who felt like outsiders on Christmas Eve and Christmas Day," says the rabbi. "If you go to a Chinese restaurant, you become an insider. You can celebrate somebody else's birthday and yet be amongst friends and family and members of the tribe, thereby the outsider on Christmas becomes the insider."

My parents always made a reservation at Wong Kee, because they weren't (still aren't) the "Let's see what happens!" types, even though the place was a hole-in-the-wall and I'm not even sure they took reservations. But when we'd arrive, an order of duck breast appetizer would be waiting for us at our table. Every time. This really impressed me—walking into a restaurant to find *our favorite appetizer* waiting for us. All Jewish families have their Chinese food order from which they very rarely deviate—ours was wonton soup, vegetable lo mein, moo-shu pork crepes, char siu, and cold boiled duck breast with scallion ginger sauce as an appetizer.

That appetizer placed on the table before we arrived said, *We know you, we appreciate you, and thank you for coming*, in the most delicious way.

It's my belief that above and beyond a great concept and execution, good luck, and good timing, it's this kind of considered service that makes the difference—the third rail of love that can be as thrilling to find in a hole-in-the-wall in Chinatown as at a world-renowned restaurant.

In 2014, after a Varnish pop-up in Stockholm, my GM Max and I stopped to eat lunch at NOMA, in Copenhagen—the two-Michelin-star restaurant (often called the best restaurant in the world) from chef René Redzepi. Our reservation had been made online two months in advance, and while we were hoping for dinner, lunch was the only slot we could get. I knew a number of chefs who could have reached out to help make the dinner happen, but we weren't looking to create a scene; Max and I just wanted to experience all the food and service wonders we'd heard so much about: the nonhierarchical staff, the dishes that looked like paintings, how fermentation translated to a cuisine . . .

Copenhagen in the fall is overcast and drizzly. Max and I walked down the cobblestoned streets following the harbor until we found ourselves standing in front of a former herring warehouse. I felt like I was in a Lars von Trier film.

Before going in, we flipped around and took a selfie.

You gotta do it when you gotta do it.

The second we stepped inside, we were greeted by the host and a dozen members of the kitchen and service staff, as if they had seen us coming. It was wonderfully imposing and caught us completely off guard. Literally tipped us on our heels. It felt like when I was given a bicycle for my fifth birthday, which was so awesome but so intense, I cried.

At NOMA, they call this "the big hello," and while I didn't cry, I did feel overwhelmed. "It's so important to look your guests in the eye, and tell them 'welcome,'" Redzepi has said about the tradition. "We can have the best menu in the world, but what makes it special is human contact."

"Eric and Max, welcome," said the maître d'. "You're the gentlemen from L.A. that run The Varnish, yes?" We had made no mention of The Varnish in our reservation. We hadn't even introduced ourselves or spoken a word. But somehow, they knew. I pictured the reservationist calling up our dossier weeks ago from an underground vault, like a non-murder-y scene in a James Bond film.

The room, with just twelve tables, felt incredibly intimate, but somehow also light and spacious. Floor-to-ceiling arched windows that ran across both exterior walls flooded the space with natural light, weathered wooden beams held up the ceiling, and wide wooden planks that felt like they'd been milled from a single, massive tree made up the floor. The table we sat down at was round, and the chairs had throws draped over their backs.

It was all very *Game of Thrones*.

We'd read a ton about the place, but there was no way to anticipate the attention to detail: the food that didn't look like food served in tiny wooden bowls, the sea rocks that dressed the plates and wild microgreens pulled straight from the ocean. We knew everything was cooked with what they could "find underground, above ground, near the water and in the trees," that they had "foraged for the freshest ingredients in every lake, river, stream, meadow and woodland," but Max and I had grown so weary of catchphrase menu descriptors like "handpicked," "hyperlocal," and "artisanal," they'd ceased to mean anything.

But this was different. This *tasted* different. This *looked* different. Above all else, this *felt* different. We had a main server, but every course was brought out by someone new, and that someone new had just cooked the food they were presenting. Being served in this manner at NOMA reinforced the point that no one position is more important than another. That a job doesn't have to be defined by a script, and to be "served" doesn't mean to oblige, but to share.

After our meal, we were given a tour of the kitchen, and after that, James, the maître d', slipped us out the back door and upstairs to René's food lab. There, we were invited to join all fifty staff members for their meal of Brazilian moqueca stew—there are so many different cultures working at NOMA that every night a different kitchen member prepares staff meal as a way of sharing where they're from.

We had just eaten a twenty-five-course lunch so we did the best we could, wanting only to experience it all, talk to everyone, see everything . . . And after *that*, we were escorted to a lounge off the restaurant for cookies, coffee, a digestif, and the biggest bill I have ever paid for a single meal.

What has stayed with me most from the experience, above and beyond the food and the room and the wine and the price, is the way we were welcomed into their world; regardless of the fact that we lived thousands of miles away and worked in bars instead of restaurants, we left feeling like members of the tribe.

CHAPTER THIRTEEN

OCD/ABK

Withered garnishes. Votives about to expire. Shreds of napkin from a nervous guest. A piece of ginger candy crushing underfoot. Fossilized lipstick rims a freshly poured daiquiri. The music is so low I can hear ice exploding in a tin when it's only quarter of the way cooked. At the bar, a cocktail glass sweats, dripping onto the leather rail. Order pads, menus, and credit cards are stacked at the service station as if members of the same genus, and the lights aren't dimmed to the appropriate level, despite the silver Sharpie markings on the switch panel. The lights—the fucking lights drive me mad. Sari sneaks a peek at her cell phone. "But I was just checking a cocktail spec, I swear," she says. Before I can reply, my focus is pulled toward a girl sitting at a booth . . . *Is she chewing gum? Is her jacket wadded up in the corner like a dead possum? Are her friends stabbing angrily at their iPhones because they can't get reception?* I can just see her apartment: unwashed dishes towering in the kitchen sink, clothes in piles hither and yon, toothpaste squeezed from the center, crusted on the sides, goo at the opening, the top lost, fallen down the drain long ago. My server Aly walks toward me carrying a full tray of empty glasses with a lit candle in the center. I lean over and blow out the torch. She looks at me like a teenager caught masturbating. "You know the drill," I say.

The drill: A tray of fresh drinks should be lit by a candle. A tray of dirties, whisked away as inconspicuously as possible *without* a candle. It's psychological. When someone sees an illuminated tray of consumed drinks, glasses lipstick-blemished and fingerprint-stained, it triggers the neural pathways in their brain that the party's over.* But if all you see is tray after tray of lovingly lit bouquets of cocktails floated in the hands of a server in an endless procession around the room, the party rages on.

Aly smiles apologetically as I pull open the washroom door and she disappears inside. Before I can accept her digressions, I'm rooted to my spot by the sound of a bottle crashing into a recycling bin. My nerves vibrate like that time my brother convinced me to bite into a piece of aluminum foil, promising, "It's going to feel *amazing.*"

With the exception of a Jewish wedding, there is no reason to smash glass. You can simply place it gently into the garbage. In terms of short-term efficiency, sure, dropping a bottle as you pass by the bin without stopping is faster, but in the long run it's more efficient to take a moment to ensure the shattering sounds of glass don't make everyone in the bar turn around and go, "What the fuck was that?!" and vow to never return.

And so, like most times when someone accuses us of "acting fancy," we're actually doing something purposeful when we place bottles gently in the garbage, when we don't illuminate detritus with a lit candle, and when we hold and serve cocktails from the lowest point on the glass with two caliper-like fingers—not to look posh, but to keep our 98-degree hands from warming the glass. If

---

* Not scientifically confirmed.

we grabbed the coupe by its body, the florid pads of our fingertips would erase the snow-like condensation coating the outside, that beautiful white crackle that indicates ultimate chill.

In the real world, people accuse me of being obsessive-compulsive, but in the bar world, my disorder goes by the name "systems and order"—pieces of a puzzle that together form a unified survival code that allows us to achieve the highest level of service possible. The key to this is knolling.

In 1987, the architect Frank Gehry was working on creating bent laminate chairs for Florence Knoll furniture, twisting pieces of maple wood into what looked like free-form woven baskets that had been taken apart and then half-reassembled and which apparently made for super-comfortable seating. His studio janitor, Andrew Kromelow, would come in after everyone had left work for the day and clean, and after he was done cleaning would organize all the workers' haphazardly placed tools and writing implements at 90-degree angles on their desks, grouping them together according to their use.

He called what he did *knolling*.

"Everyone always looked forward to coming into work in the morning because the space was clean and well organized," Andrew told me. "Whether they were office workers, carpenters, or technicians, their environment was ready for them to work."

Andrew's obsessive reckoning may have gone unheralded outside the studio's four walls were it not for his friend and co-worker at the studio Tom Sachs. Tom incorporated knolling into his creative work, and when he opened his own studio years later, adopted the phrase "Always be knolling," or ABK, as his mantra.* Since then,

---

* A derivation of the line "Always be closing," from David Mamet's *Glengarry Glen Ross*.

knolling has inspired art projects, a blog called *Things Organized Neatly*, books, maps, magazine layouts, and an Instagram account with 280,000 followers dedicated to "flatlays," which are things organized and laid out flat, which creates incredibly pleasing optics.

"I am proud to have brought knolling to the world," Andrew told me in reference to being the inspiration for a cult of obsessive-compulsive reckoning. "I hope the next generation takes it one step further."

AT THE VARNISH, EVERYTHING is knolled. Behind the bar, every tool, bottle, pen, check presenter, and glass has a home. When everything lives in the same place all the time, any one person can step in and work, their movements as precise and economical as the last person, because they've been trained using the same system.

In other professions, knolling goes by the name "lean"—an efficiency process that reduces waste in materials, time, and labor. It is said to have its origins in Venice in 1104 at the Arsenal—a complex of shipyards created to build warships for the Venetian navy. In order to build hundreds of galleys every year (large seagoing vessels propelled primarily by oars), the builders devised a standardized design: the hull was completed first, then they'd flow it down to the next assembly point, and on and on down the line, where at each stop, every item needed to complete the ship was added. By 1574, the Arsenal's practices were so advanced that King Henry III of France was invited to watch the construction, which from start to finish took place in less than an hour.

This system was adopted and improved upon for more than three centuries, until most notably, Henry Ford combined it with technological advances in 1914, which allowed him to produce a Model T in two and a half hours. Approximately twenty years later,

Kiichiro Toyoda and Taiichi Ohno upped the ante by inventing the Toyota Production System—a now famous system that improved Toyota's output times and lags between process steps, which led to a better-quality, lower-cost car, with the shortest-ever lead time.

In 1990, MIT published *The Machine That Changed the World*, which stated that the Toyota Production System was so effective and efficient it represented a completely new paradigm. They called this paradigm "lean production."[*]

Applying principles of lean and knolling at The Varnish is what makes the place tick. From the seen to the unseen, every element helps our flow, eliminates redundancies, and creates a seamless experience for our staff and guests. In a fast-paced environment like a bar, every move either adds to your flow or pushes you further into the weeds. So it's imperative that everyone work with the same sense of urgency. Every step of service affects every *other* step of service, which everyone working knows, since everyone works every position.

Staggering is key.

If I were hosting and had two parties waiting to be seated, I would seat one party, then return to the host stand and wait until I saw the server greeting them before seating the other party. This gives the server a cushion between taking their first and second table's order, which gives the bar a cushion between receiving them. If instead I seated both parties at once, and the server had to attend to both tables at once, they would then drop off both tables' drink tickets to the bar at the same time—anywhere from four to twelve drinks—which can quickly bottleneck service.

When the server is ready to take a table's order, they write the drinks down on a chit, using position numbers like at a restaurant:

---

[*] A woefully selective and abbreviated history of "lean."

facing the table and counting clockwise, the first customer to the left is position 1, the next is position 2, the next 3, and so on. This way, anyone delivering the order to the table will be able to give the right drink to the right person. This chit is then handed to the bartender to make the drinks, and once the drinks have been trayed, the server ensures they're placed clockwise on the tray in the right positions as they garnish each one. Before leaving the wait station, they quickly tap into their third eye, or "bar eye," identifying every potential situation that needs attention on the floor and setting up for it before leaving: Table 7 has left and needs clearing, so the server stuffs a bar rag in their right back pocket. They notice there's a line at the door, so they stack a few napkins onto the tray so they can drop them when the guests are seated. A gentleman at table 12 is holding out his check presenter with a credit card—mental note to grab that after dropping the drinks and before clearing table 7.

Go time! Lift the tray of drinks off the pizza stand and pull out the one sitting on the lower rung. Pop it onto the top shelf so the bartender can tray the next round. Then it's on to the floor to drop the trayed drinks at table 3, chat chat, smile smile, answer questions, and then a quick half-circle to the other side of the piano to grab table 12's check. On the way back to the middle of the bar, place napkins down for the newly seated guests at table 9—"I'll be back in a few to take your order"—and then move two tables backward to clear 7 of empty glasses and used napkins onto the now empty tray, while blowing out the candle so it's not all lit up. That table gets a wipe with the bar rag and then it's back to the dish room, going the long way to pass Anthony, the host at the front, to tell him table 7 is clean and ready to be seated.

Every single step on the floor has to be done in order of what's most important and how it can create the most continuous flow. The number of steps and amount of time each turn around a room takes affects every other position's flow and, ultimately, the guests' experience.

After dropping the tray of dirties from table 7 with Carlos who will bus it and put it back into service, the server runs table 12's check at the dimmed POS system and, as it's spitting out the receipt, tweaks the garnishes on the tray up for table 6. Clip the receipt to the top of the check presenter with the card, grab extra napkins, ('cause you never know), pick up the trayed drinks, move the bottom tray to the top, and head back out to the floor, dropping the drinks, then the check, and heading to table 9 to take their order.

If first impressions matter, last impressions are what a guest leaves the bar with, and when a guest wants to leave, they shouldn't have to wait. Bringing the check is the "dirtiest" part of the whole experience—I want it to be quick and painless. One of the seemingly minor things that helps is using black Bic Soft Feel retractable pens. When we first opened, I bought boxes upon boxes of classic blue-and-white Bics with the caps that got lost, which made the ink dry out, which meant that when guests were ready to sign their checks they were going *scratch scratch scratch*, with nothing coming out. They were shaking their heads in consternation. Trying again—*scratch scratch scratch*. Sighing. Waving their hands to get the server's attention. That server had to pivot from their meticulously designed route that didn't waste steps to go over to the table to see what was wrong and then backtrack—grab a new pen at the service station and scribble on a pad to make sure it worked before bringing it over to the guests and apologizing and then returning to what they were doing before.

Sure, this isn't a tragedy, but imagine it happening over and over again. Which it does. Like a Rube Goldberg contraption where one dropped ball affects things in the most indirect and complicated of ways, it throws off the flow of the night.

Not to go on and on about pens—and you can see how this isn't even about pens, but about figuring out the most efficient way to execute—when those pens weren't drying out, they were secreting all their ink in an unseen spot inside their cavity so when guests tried to use them, no ink came out the tip but, instead, all over their hand.

This is not an advertisement for Bic Soft Feel retractable pens, but with them we have no more lost caps. No more dried-out cartridges. No more leaky ink!

End of pen rant.

Then there's all the unseen stuff, which is as important as the seen for elevating our guests' experience. A Perlick glass froster that sticks when a bartender opens it (constantly) can be a serious shift killer. "Open . . . you . . . fucking . . . fuck!" is not an uncommon bartender's cry when forced to stop midflow and coerce the froster to open. Silicone lubricant makes a frequent appearance on my Home Depot list to keep this from happening.

Diligent turning on of the thermostat is a lesson learned the hard way. The Varnish, as mentioned, is not only a small, windowless closet, but sandwiched between a kitchen and Skid Row, and if we don't turn our thermostat to 71 degrees at least four hours before service, the room will never catch up, never be cool enough, and everyone will fry. One day, three weeks after opening, we did not remember to turn our thermostat to 71, and it was summer. And it was fucking hot. And because it was summer and fucking hot and we hadn't remembered to turn our thermostat to 71, our HVAC broke. You could blow on somebody's hair with all your might and it wouldn't move, just lie like a damp rag down their back. And because it was a Saturday night and no HVAC servicemen were around until Monday or maybe Tuesday, we had no choice but to turn the bar into *Havana Nights*. Someone went down to Santee Alley and grabbed a stack of guayaberas for the staff and a few extras for our regulars, and for the rest of the weekend we worked in lightweight pocketed shirts, piped Buena Vista Social Club through the speakers, served mojitos and Cuba Libres, brought in a big oscillating fan, and offered handheld folding fans to all the tables. I felt like I was back in Cuba with my brother, Jean Michel, and our friend Ben for my brother's bachelor party, drinking shoulder to shoulder in the tiny La Bodeguita del Medio bar where Hemingway drank his mojitos.

KNOLLING AND LEAN ARE NOT static systems; they live and breathe through every action, every employee, and continuous repetition as we do our best to "manage toward perfection." At night, our last order of reckoning is done when we lay our clean gear out on top of the bar and ensure that for both stations there are two muddlers, eight stirring spoons, four cracking spoons, twelve graduated jiggers, two pairs of tongs, and thirty shaking tin sets, setting up whoever's coming in next for success. I do the same thing with my adventure gear before an expedition—lay it out across my living room floor to check that I have all my survival items and that they're in working condition: hiking boots, extra shoelaces, headlamp, backup batteries, freeze-dried meals, water purification tablets, first aid kit, gloves and glove liners, socks and sock liners, and duct tape.

My love for adventure and knolling, even though I didn't know what it was back then, is another thing that's been ingrained in me since childhood. Between the ages of eleven and seventeen, I spent my summers racing sailboats. Blue Jays at fourteen feet with a mainsail, jib, and spinnaker. On a boat, everything is knolled, and everyone works with the utmost efficiency. It's the only way to survive both no-wind and high-wind conditions where, in each instance, precise movements and actions are required to cut through the water.

In my midtwenties, I fished for salmon in Prince William Sound off the southern coast of Alaska on a fifty-five-foot diesel purse seiner* where the backbreaking labor was all about ABK—no Grundéns† belowdecks, but always removed and hung on the rack just outside. All ropes were coiled clockwise; otherwise the boat might "tip" from inconsistency. Corks, or floats, at the top of the seine had to be organized and untangled, ready for the next set and then the next and then the next—up to six times per day, six days a week, occasionally seven, for two months.

When we moved on to halibut fishing, things were less of a sprint and more of a marathon—buckets of hundred-foot line with hooks every ten feet needing to be baited and coiled into large buckets. This type of fishing is called longlining,‡ and the meditation of baiting and setting buckets out to sea is a real test of patience. Perhaps an even greater test is the time spent untangling knots caused by fighting sand sharks, rockfish, and the roiling boil of rough waters.

Those summers I spent learning the intricacies of ships' systems were the foundation for all the high-intensity stuff I love today: riding my Ducati Monster S2R through the San Gabriel Mountains on Angeles Crest, skydiving over San Diego, and hiking mountains with altitudes of over fourteen thousand feet—Kilimanjaro, Mount Rainier, and El Pico de Orizaba. Every daredevil activity demands I have systems and processes in place, intense training

---

* A fishing boat with a type of net called a seine that hangs vertically in the water with its bottom edge held down by weights and its top edge buoyed by floats.

† Waterproof commercial fishing gear worn at all times on deck to stay dry.

‡ A commercial fishing technique where the "long" or main line is baited at various intervals via a branch line or "snood," which the hook is attached to.

and preparation, and a well-considered set of potential outcomes with gear and medical supplies packed to mediate. Which is exactly how a bar should be run—like a life-or-death adventure.

Every night at The Varnish we try to consider what might go wrong and how we might remedy it. Of course it's fantasy to imagine you can anticipate half the shit that's going to happen, but if the bar is knolled, if staff are on the same page about all the stuff they have to be on the same page about, if everyone is trained and focused and all in this thing together, we can survive the night.

CHAPTER FOURTEEN

VICE

It's too bright to be my bedroom. The pillow's too flat. It smells like a drum circle. My eyes peel cautiously open and blink. Catch the curve of a girl's back rising and falling in a fitful heave.

What's her name? . . .

What's her name? . . .

What the fuck is her name? . . .

I swing out of bed and beeline for the toilet, or rather, beeline for where I imagine the toilet to be, and by "beeline" I mean hobble on one foot since I can't find my crutches—a tragic handicap resulting from last month's off-road motorcycle accident in Stanislaus National Forest. The hallway leads to a kitchen, which I have vague memories of stumbling through last night. A long, dark counter bows under the weight of a Vitamix surrounded by an apothecary of jars: Sex Dust, Sun Potion, Soylent Powder, Moon Juice, Balls in the Air, He Shou Wu, Hemp, Maca, Baobab, Camu, Lion's Mane Mushroom Elixir, Nutritional Yeast, Jesus Christ I have to piss.

I slide open a reclaimed-wood barn door which thankfully deposits me into the bathroom. Once inside, I sit down on the toilet to gather my thoughts. For anyone in the hospitality industry, bathrooms are a refuge. With everyone at your throat, at arm's length, all night, bathrooms—palatial or shithole—are yours and yours alone. Unless of course you invite someone in with you, and then . . . Last night comes back in a rush: Me. Her. Last call. The middle bathroom at The Otheroom. What the hell compelled me to take her into the bathroom to fool around, because have you ever been in a bar bathroom at 2 a.m.? It isn't like a house party bathroom, or a restaurant bathroom. It isn't even like a gas station bathroom, though that may be its closest kin. A bar bathroom is the bathroom of people without aim—toilet paper, tampons, condoms, piss . . . Murder scenes aside, it might be the most repulsive place to have sex.

I flush and stand up to wash my hands, the mirror over the toilet confirming I've barely slept. I wipe my hands on an organic waffle hand towel and hobble back through the witchy kitchen into the bedroom, where I see, through the window, the horizon dropping into the ocean.

Fuck. I'm on the Westside.

I grab a purple glitter pen off the nightstand and on a notepad adorned with the aspirational quote YOU ARE SO LOVED!, write, "Thanks!" Do my phone/keys/wallet check and crutch out as quietly as I can.

I LOOK TO THE EAST. I look to the west. Nope. No idea where I left my car.

I text Bostick: *Yo! What happened last night after Copa? Did we drive? Do you know where I parked?*

I hit SEND and notice the time: 8:47 a.m. No fucking way Bostick's up anytime soon. I squint through the glaring sun, wishing I had my shades, which are in my car. I slip my phone into my back pocket, wiggle my Sheepette crutch covers under my armpits, and head west on San Juan.

For a while, that whole not-sleeping-around plan was going really well. I went out on a couple dates that didn't take place in bars, including a free jazz picnic in the park by the La Brea Tar Pits, a bike ride on the Santa Monica Boardwalk, and a hike up Bronson Canyon which almost killed her dachshund, so I picked the little wiener up and carried him in my backpack. And now I've gone and fucked up my adulting streak by sleeping with a woman whose name I still can't remember but whose face I know well. She's not an every-night-of-the-week regular or one of those crazy-town weekend warriors, but an every-once-in-a-blue-moon-happy-to-see-you kind of regular who loves The Varnish enough to visit when she's on the Eastside, and now shit's gonna be weird when she visits. It's pretty awkward asking someone you had sex with for money. When you sleep with someone, and they visit you at work, and that work is a bar, you're kind of obliged to let them drink for free. I'm in no shape to be dating, even though she's perfectly nice, and obviously she might be thinking the same thing, but what if she's not? What if she wants to hang out again and comes to the bar; there's nothing I can do to stop her. Unlike traditional jobs, with complicated sign-in procedures and assistants vetting who has access to you, anyone can walk into a bartender's place of business any night of the week and post up for as long as they want.

The last time I slept with a customer—we'll call her Moon—we ended up dating. But we only saw each other at the bar, because I work six or seven nights a week, which meant she was hanging out all the time. It's not like I wanted to cheat on her, but my job is pretty flirtatious, and it felt rude to flirt with others when she was there. Also, I didn't always want to see her. That might make me sound like an asshole, but imagine having no control over when you see the person you're dating. It's one thing if you work in the same office and have consciously made the decision to walk that precipitous path together; it's quite another to have no control over when your lover comes around. Finally, we had to have The Talk, during which I suggested she not hang out so much, and she told me she liked the bar better than she liked me, so we broke up. Soon thereafter, she started coming in and getting sloppy drunk, calling out for *"Shots!"* every time she saw me flirting with someone. After the third time this happened, we had to have Another Talk, during which she apologized, and I apologized, and we promised to remain friends, which for her meant coming into the bar slightly less regularly but still not infrequently and always with another guy or girl with whom she would make out. Even though I didn't love her, or at this point like her very much, it was a truly egregious situation. I was getting very close to pulling the "This is my bar, you're eighty-sixed" card when she ended up at the bar at last call, along with two other girls I'd slept with, which is a fun setup for a sitcom but not real life (see chapter 11, "Top Ten Reasons Not to Date a Bartender"), and I put her in a cab home.

That was the last time I saw her.

AT ABBOT KINNEY, I HANG a left and crutch up to The Otheroom. To the untrained eye, the bar looks like a cool, pro kind of place, but to anyone in the know it's the Bermuda Triangle of iniquity. I always end up here when I'm on the Westside.

I sit down on the redbrick ledge and text Forrest, who was bartending last night:

Me: *Hey dude.*

I hit SEND. Doubt he's awake. But seconds later:

Forrest: *Hey man.*

Me: *You're up!*

Forrest: *Haven't gone to bed.* 😂

Me: *Do you happen to know where I left my car?*

Forrest: *Nope. But when you walked in you fell over. Try Craig?*

Craig is the owner of The Otheroom and is *always awake*, except during normal business hours.

Me: *Right on. Get some sleep!*

Forrest: 👍😂

At least I didn't ask Little Miss Bathroom Stall if she needed a job, like the last time I hooked up with someone at The Otheroom. Let's call this woman Natalie. Natalie was a great bartender, but we did not make a great couple, and within weeks had decided to stop sleeping together, but by then she had quit working at The Otheroom and was now working at The Varnish. If I thought spending my night staring at a wall of women I regretted fucking was bad, I should've multiplied that by a thousand when I decided to work with someone I used to sleep with. Spending eight hours a night side by side in a booze-and-pheromone-soaked room is incredibly challenging when you're trying to consciously *uncouple*. Even though we hadn't fallen in love or broken each other's hearts, it felt shitty watching her flirt with other guys and dip into the office for the occasional recreational fun. When she actually started dating some dude, I was happy for her, but that didn't mean it wasn't difficult listening to their romantic plans and inside jokes. It made me deeply uncomfortable—okay, it fucking pissed me off—watching them ogle each other. I was jealous. It felt bad that she got over me so fast. I wanted to be back in love, or lust, or some combination thereof.

And then they broke up, and she was really sad, and I was really comforting, and instead of hopping into our respective rides and heading to our respective homes at the end of the night, we shared a ride and ended up in bed together. This kept happening over and over again. I couldn't fire her because that would be a dick move, not to mention illegal, and then one day she quit and we haven't seen each other since.

I make a mental note—*Stop fucking women you meet in your bar*—wiggle my crutches back under my sweating armpits, and head south on Abbot Kinney in search of caffeine.

INTELLIGENTSIA IS PACKED WITH POST-CLASS-GLOW yogis in skin-tight leggings and bra tops. Shirtless guys with bronzed chests. They throw me pitying looks, sad about my crutches and pale skin, the way I'm downing a double-shot espresso and pain au chocolat without considering the benefits of oolong tea and avocado toast. They don't know that until my accident six weeks ago I was one of them, reveling in my physical body. Limber. Flexible. I'm not blaming the accident on my recent bad decisions, but this lack of activity and the introduction of painkillers to my daily routine has taken its toll. When you work in hospitality, you spend hours on end doing small, repetitive movements and standing on your feet on hard floors—without some kind of physical release that moves your body in different ways, you're fucked. Whether it's running or boxing or pretzeling your body into unusual shapes as a person professing to be a sylph saunters around the room swishing sage, it's imperative to stay physically active when you work in the service industry, and without the release, I'm a mess. Unable to sweat out my anxiety or clear my mind. The only thing I have is crutching.

I check my phone—neither Bostick nor Craig has texted me back—swing my sticks back under my arms, and head to Hal's for a hair of the dog.

HAL'S IS OLD-SCHOOL VENICE BEACH. The kind of place everyone frequents—the OG artists, the newbie artists, families, surfers, drug dealers who present their wares with an orchid and a song. Families come here, surfers, all the locals, highbrow and low. I pull open the heavy door with some difficulty and am hit with the smell of fried eggs and floor cleaner. At the bar are a smattering of old-timers—could be early risers or still awake from last night. The dining room is quiet. In the back, Big Sexy's holding court at his favorite table. I slide onto a stool next to a local fossil and smile *hello*. Order a glass of rosé.

I always feel like I'm stepping through the looking glass when I find myself drinking in a bar before noon. Like some alternate version of myself stuck to his barstool drinking the same drink. Talking the same talk. A regular. I have a soft spot for regulars. Those guys (and they're mostly guys but I've known a few women) who walk into bars at 8 a.m. and drink like it's happy hour. As a bartender, it can be hard to bear witness to the slow dissolution of a regular's life. Seeing their faces every time I work can be conflicting—I'm happy they visit, because money, but I wish they'd drink a little less. Take better care of themselves. The National Institute on Alcohol Abuse and Alcoholism reports that alcohol causes 88,000 deaths each year, and while those deaths aren't on my hands, they aren't *not*. What bartenders, and by proxy, bar owners don't pay enough attention to are the customers who start rolling in every night of the week. We have a tendency to look the other way as barstools named after those who have been loved and lost start crowding the room like pigeons in a fountain. When photographs of people we used to know fill the backbar mirror and we can no longer see our reflections.

I take a sip of my wine and think about the people I've let slip through the cracks. Knowing when to say when and figuring out how much to hold myself accountable has always been a challenge. And sure, I'm not responsible for anyone else, people make their own choices, but none of us move through this life alone. Relationships we cultivate shape us, and when we see someone on the regular, we become part of their life and have the chance to influence them, for better or for worse. Like the time I had to remove Courtney's father, Fred, from the bar. Fred was a Vietnam vet with two tours as a helicopter pilot under his belt, and a third with Air America.* Since the end of the war he'd been drinking and drugging away his undiagnosed PTSD. Courtney and I had only been in L.A. for six months when we flew down to Houston, where Fred was living, for an intervention. He hadn't been answering Courtney's calls. We arrived at his house to find him naked in his bathtub, submerged in days-old water, muttering racial slurs under his breath. When we tried to move him he screamed, "You're blowing my cover!" A cornucopia of prescription bottles and primary-colored pills crunched underfoot as we carried his ninety-pound frame into our rental car and on to the VA hospital.

A week later, we moved Fred to L.A., where he slept on our couch, furtively smoking cigarettes, drinking, and arguing with infomercials in his tighty-whities, and when he got bored he'd wander down to The Varnish. He was particularly fond of Monte Carlos—an old-fashioned variation—and commandeered a booth in the early

---

* Air America was operated by the CIA during the Vietnam War and supported covert operations.

evenings, where he'd sit and glare at everyone who walked in. He could be charming and composed or mean and sloppy, but it was impossible to know which way he would tilt until it was too late.

One night, he got into a heated argument about the movie *Apocalypse Now* with one of our regulars, Ken, a sweet Asian gentleman. They were talking about the knife that Captain Willard uses to kill Kurtz. Ken was sure it was a bolo, which looks like a short machete, and Fred was sure it was a Special Forces SOG like the one he always carried. To prove his point, Fred pulled his knife out of his hip sheath, his eyes slightly glazed. Ken went quiet as Fred said, "It's like this one. You see?" He brandished it in front of Ken's face. "This is the knife he used!"

Nobody likes a knife fight, especially a bar owner, and when it's your potential future father-in-law that's even worse.

"Fred, you're right," I told him. "Let's put the knife away."

"Your friend doesn't believe me!" he yelled.

I turned to Ken and said with raised eyebrows, "Yes he does. He sure does. Right Ken?"

Ken nodded.

"Let's take a walk," I said, quietly but firmly.

"I want to finish my . . . What's that drink I like? Tell me again, I keep forgetting."

"A Monte Carlo. Let's make you one and take it with us to the loft."

"Yeah!" he exclaimed. "That's the one. Good plan." Fred placed his knife back in its sheath, grabbed his drink, and walked with me out the back door.

"ANOTHER ROSÉ?" BILLY, HAL'S BARTENDER, asks me.

"Ah, no thanks," I say, and glance at my phone, noting the bars are dangerously low and no one, still, has texted me back. "One for this guy though," I nod to the old-timer next to me, who smiles his gummy thanks. "And the check."

BACK OUTSIDE, ABBOT KINNEY IS coming alive like a fucking musical: wind chimes made out of seashells and glass are clanging in the ocean breeze. Skateboards are schussing down the street, helmed by sun-kissed boys in boardshorts and freckle-faced girls in bikini tops. Everyone looks like they're sixteen. I wonder if I'd be a different person if I'd grown up here. If I'd have hung out at a surf shop instead of telling my parents I was studying at my best friend Ygael's house when really we were driving into the city to eat acid at Limelight—the Church of the Holy Communion on the corner of Twentieth Street and Sixth Avenue that turned into a nightclub. Thanks to Mary Beth and Muffy Foley, who were seventeen and hot and wore absolutely fabulous outfits like the girls in the movie *Heathers*, we got VIP treatment. We always cut the line, managers plied us with drink cards, and the girls hung out in the DJ booth with the Limelight's eye-patch-wearing owner, Peter Gatien. Sometimes Mary Beth danced in the steel cage under the church's fifty-three-foot-high apse like a Madonna video as we danced underneath. When she was lowered back down to the floor, the security team would make sure I was standing by to escort her out.

Pretty dope for a sixteen-year-old.

It was right around then that I became an entrepreneur. A Deadhead at a show told me and my friends Doug and Paul about a nitrous gas facility that had no security aside from a barbed wire fence, and just beyond that, tons of tanks lying around. A few scars and stitches later, we were selling balloons of the stuff in concert parking lots for $5 a pop. One night Doug, Paul, and I got pulled over on our way to meet some friends who were hiding out in a nature conservatory and were promptly arrested for possession.

It was impossible to miss the shiny silver tank in the back of the Corolla. We called Ygael from prison and he used the $1,500 he'd saved up to buy Chris Elam's pink 1962 Mercury Monterey, which we were all going to prom in, to bail us out. Paul had a prior trespass, so he spent the summer doing community service. But because we were all minors, and it was our first offense, the charges were dismissed and our records were expunged. Doug and I, indebted to Ygael, took jobs at his family's French delicatessen, Délices. Doug worked the counter, and I ran the floor at lunch. My first job in the service industry.

MY SECOND AND THIRD ARRESTS were for possession of weed (sorry Mom and Dad). Kids don't have to worry about that today those lucky fucks. Overall, though, I've had a pretty okay relationship with drugs. I think it has something to do with my OCD. Or my role as the oldest son. But I can go off the rails at after-hours and lock-ins at the bar. I find it hard to resist those nights when bartenders, brand reps, deep regulars, and dealers huddle around tables and smash up against each other in booths as Zeppelin crackles over speakers meant for jazz. When we breaststroke through clouds of cigarette smoke, roaming from table to table to the bar and back to the office. Any "couple" hanging out in a booth are "invisible." If the office is locked, don't bother knocking. Lines are divided along the bar top's drink rail, a stone's throw from Shaking Tin Jenga—a game that involves stacking bar tins in a pyramid, then dismantling it without toppling it. Each successful stack+unstack, you get a toot.

Sometimes Carlos stays on, sipping orange juice and laughing with everyone as he cleans. If somebody tries to help he says, "No, please sit."

Only when we've used up all our *I love you mans* on each other do we throw down a pile of cash for Carlos by way of an apology and creep out the side door.

During one of these nights last year, Death & Co proprietor Dave Kaplan came up with the idea to take the lock-ins out for what he called "L.A.'s Down & Dirty Tour." The inaugural night, eighteen people met at The Varnish for an aperitivo, then hopped into three limos stockpiled with all of the drugs, pills, powders, and bottled water a battery of bartenders could ask for. First stop was the Body Shop, followed by Gold Diggers, and then Sam's Hofbrau. We drank zombies at Tiki-Ti, on Sunset, then stumbled next door to El Chavo for after-hours, having rented out the three-room hotel upstairs. The entire night was "sponsored" by a rep, meaning someone convinced some brand that picking up the bill for a night of the city's "top bartenders" celebrating "Los Angeles bar culture" would be a good look.

Down & Dirty was an epic version of what quickly became a regular occurrence, because in this industry, there's always a party happening somewhere.

"Yo, Simon's in town for one night only, you gotta come out."

Or: "I'm leaving for New York for a week, come send me off."

Or: "It's my birthday dude, don't fuck it up."

Or: "I'm bartending tonight come visit."

Or: "Topaz at Jumbo's asked about you . . . oh yeah, we're here."

Whatever you're into, if you meet people doing it, that's going to become your crew. Like if I went to temple, I'd make a bunch of friends who went to temple, and even when I wanted to spend my Friday night at the movies instead of in shul, I'd end up reading the Talmud because that's what all my friends were doing.

But I do not go to temple (sorry Dad); I go to bars, and I meet men and women who go to bars, and when I don't want to go to bars, it's tough to be Debbie Downer not living up to everyone's expectations. And this, my friends, is the definition of "a vicious cycle"—best illustrated by Antoine de Saint-Exupéry.[*]

"Why are you drinking?" the little prince asked.

"In order to forget," replied the drunkard.

"To forget what?" inquired the little prince, who was already feeling sorry for him.

"To forget that I am ashamed," the drunkard confessed, hanging his head.

"Ashamed of what?" asked the little prince who wanted to help him.

"Ashamed of drinking!" concluded the drunkard, withdrawing into total silence.

My pocket buzzes.

It's a text from Bostick!: *Damn. I feel rode hard and put away wet.*

Me: *Dude, where's my car?*

Bostick: *You left it in a lot by Copa. Last I remember is you doing tequila shots at The Otheroom from their stash bottle and bumps in the middle bathroom. See you tonight at The V. I'm in at 8!*

The lot off Second Street in the alley by Copa d'Oro is empty, save for a sign stating any cars left after 7 a.m. will be towed to Tip Top Tow.

It is now 11 a.m.

I call a cab, sit down on a fire hydrant, and watch the Santa Monica Ferris Wheel spin around in lazy, slow circles. The good news is that I didn't drink and drug and drive. The bad news is that I'm embarrassed I thought I had. The last time I did something that

---

[*] From his novella *The Little Prince.*

stupid was three years ago, when I took Courtney's brother, John, to a Dodgers–Rockies game for his twenty-third birthday. John grew up with his mom in Boulder and is a huge Rockies fan, the kind of guy who paints his face and grabs Jumbotron attention. We started our night with beers and shots at Gold Room—an old haunt with blue and red strips of neon lighting, a carpeted floor, and a faux stone bar top, aka the perfect place to pregame. John was dressed in full Rockies regalia (thankfully he left the face paint at home), but we were still not a popular duo, L.A. being filled with hard-core Dodger fans who "bleed blue." We left the bar, as well as my car parked on Sunset, and walked up to Elysian Park, dropping into the Short Stop for another round of beers and shots, and when we left *there*, bought some beers at a bodega, sheathed them in brown paper bags, and made our way up Vin Scully Avenue to Stadium Way, with John hollering, "Let's go, Rockies!"

I'd bought really good seats in a VIP field box in section 21, along the third base line, where we continued to drink and devour Dodger Dogs and Dodger Nachos Helmets filled with cheese, refried beans, sour cream, pico de gallo, and carne asada. As it turned out, we were sitting a few rows behind Jason Bateman, and at one point, in the middle of the eighth inning with the Dodgers losing, they played "Don't Stop Believin'" and the cameras were positioned on Bateman, with me and John air-guitaring behind him on the Jumbotron.

Epic!

The Rockies won, and we somehow made it out of there without getting our asses kicked. I was having a helluva time showing my girlfriend's brother that I could hang. That he was like a brother to me. That I was the cool boyfriend.

We stumbled back to the Short Stop for Crown Royal, in honor of his pops, Fred, who swears by the stuff, and then hopped into my car like it was nothing. Driving John seemed like the responsible thing to do, considering I was his chaperone and wanted him to get home safe. We blasted the radio and sailed through the night, street and traffic lights like flares through a camera lens in my eyes.

Outside John's house, I shook him awake, took him inside, then got back into my Nissan and headed home. Courtney was awake when I walked into the bedroom and plopped myself down on the bed, except I miscalculated and fell onto the floor.

"So the game was good?" Courtney asked, her voice quiet and low.

"Sure was," I laughed.

"And John had a good time?"

"Sure did!"

"And how did he get home?"

"I drove him!" I told her, pulling myself up.

"You *drove him*?"

"Yup."

She moved slightly away from me when I managed to make it onto the bed. "You drove him home and then you drove here?"

"Yeahhh. That's what happened. That's what I said."

"Don't ever do that again," Courtney told me slowly. "Don't ever put two lives I love at risk. And definitely don't ever show my brother that's the thing to do."

She walked out of the room. A minute later, I heard the front door slam. The next morning, I woke up alone and fully clothed.

"THAT'LL BE $262," THE GUY behind the chicken wire at Tip Top Tow tells me.

I hand over the money and in exchange I'm given my keys. I feel like I just won a carnival prize after too many tries. I throw my crutches into the back and fall into the driver's seat. Everything is boiling hot. My own private hell. I take the side streets back home because I can't handle the freeway, and when I make a left on Gower, catch a glimpse of the Hollywood sign. In all my time in L.A. I've never hiked to it. My actor pals are always yammering on about whether it's a good day for the Hollyridge Trail, or maybe they're feeling like something a little longer, so "Yeah we're gonna take Innsdale Drive," or if they're getting in shape for a role, they brave Brush Canyon Trail, which is 6.4 miles round-trip with a 1,050-foot elevation change. If I was an actor I'd have more time for daytime hikes. Regret is a waste of time, and envy too. But I've felt both at times hanging out with my peers, who instead of falling into the bar business stuck it out as interns, PAs, and executive assistants, working for shitty pay in lowly positions, and are now successful actors, writers, directors, and producers. While I worked in bars, they cultivated relationships in their industry and year after year, moved collectively up the food chain. I still get a pull in the pit of my stomach sometimes wondering *what if?* You'd think it'd be assuaged the times I get cast in independent films, playing supporting roles or doing stunts because my good friend Joey Box is a stunt coordinator. And by "doing stunts" I mean playing an innocent bystander who gets knocked over the head by the lead in a TV show. But every time I start feeling insecure about my choices, I have a genuine moment with a guest or a staff member or solve a problem that makes me feel great. I walk into The Varnish and feel proud of what I've built and the jobs I've created. I feel stable and calm—two sensations I never knew as an actor. I know my actor friends struggle on their end too—that it's not all as wonderful as it looks.

The sign fades in my rearview mirror, and I know the game of guessing what might have been is pointless. We don't have time machines. We have choices. Sometimes we make good ones and sometimes we make bad ones. Sometimes we hike up mountains and sometimes we slide all the way the fuck down. Today I woke up in a bungalow in Venice Beach. Who knows what tomorrow will bring?

CHAPTER FIFTEEN

CLOSING TIME

It's 1:30 a.m. Saturday. The booths are full. There's a crowd two deep at the bar. The warehouse lights are throwing a hazy glow over the room, as if our resident horror film director Jeremy Kasten, high on mushrooms and wandering from table to table randomly choosing strangers to sit down with, swapped them out when no one was looking.

"Hey man," I rest on my haunches next to him so that neither he nor the people at the table feel like I'm intruding, "we're working on a cocktail that could use your palate."

This is a lie. We definitely don't need Jeremy's palate, which can't taste anything, which is why his drinks are going down so fast.

Jeremy slides up to a standing position and graciously apologizes to the table for leaving. Tells them he is needed elsewhere. "But I'll be back!" he yells over his shoulder.

I prop him up by the service station and hand him a daiquiri that sat too long and died.

"Oooh I *love it*," he says with a smile.

"Cool buddy," I tell him. "Thanks for the feedback."

Rebekka swooshes past me in her 1940s poodle skirt and tosses a chit to Devon, who's slammed at the service station with six tins on the scupper and three on the board. The charismatic celebrity who's become something of a regular is perched on the piano bench with Jamie, playing and singing to the thrill of everyone in the room. I'm gonna have to help him escape through the back to avoid paparazzi.

"Hey there East Coast player!" Kimmy crows to Richie behind the bar. "So sweet to see you back there."

Richie, who's visiting from New York, hopped into the personality well a little over an hour ago after Eugene got cut at midnight when it looked like things were slowing down, which was a rookie move because *the minute* you cut someone you get a rush. At the same time I was cutting Eugene, the Edison let out early thanks to a power outage (ironic), and everyone poured in here.

"I wanted to make it over tonight, and by the grace of LADWP my prayers were answered," says Joe Brooke.

Joe slings at the Edison, and we rarely see each other except at industry events.

"Always happy to see you buddy. We doing the next Sporting Life* at your spot?"

"I hope so," he nods and looks over at Richie. "And now that I see him back there, I might find a moment to ask him if he'd join. Do you think he'd be into it?"

Joe is from a small town in New York that bumps up against the New Jersey state line. He's too humble and sweet as hell, which I hope never changes.

"Ask him once we finish last call," I tell him.

"Okay. Cool. When is that?"

---

* A monthly industry gathering to chat, hang, and drink that started in the back of Bar Keeper in Silver Lake, then spread out across L.A., taking place in a different venue every time.

I peek at my iPhone. "Oh shit. Now."

I cue "Je T'Aime . . . Moi Non Plus," by Jane Birkin and Serge Gainsbourg—one of Sasha's favorites—and raise the lights midway up the track, past the fading Sharpie marks, which is a subtle cue the night is coming to a close. We never throw the fuck-off lights up right away; these are more like the "We love you and thank you, but fuck off pretty please?" lights.

Big Jay lumbers around the room and gently reminds guests they only have a few minutes to finish up. Some tables nod and get to it. Some ignore him, which is pretty hard, since he's six foot three with hands the size of old baseball mitts. They think if they don't acknowledge his presence, they won't be asked to leave, and only when we walk over and take their drinks off the table do they exclaim, "But I wasn't finished!"

"Have a good night! Get home safe! See you again soon!" we say as everyone stumbles out, wrapped in their various states of post-drink daze.

"Do *I* have to go?" Jeremy asks with pleading eyes. "Kimmy!" he cries as she passes by. "I'm gonna marry that woman!"

"Mazel tov." She cheers his drink with one of her own and continues on to the dish room.

"You can stay, but stay out of the way all right?" I tell him and nod toward a booth, where he happily sits, sips his drink, and dreams of the mystery woman who will be his bride.

The celebrity heads out the back door with a few of the crew he came in with, leaving only the people with office privileges,[*] the staff, and the Shields brothers, Zach and Ben, who have been coming here since we opened and are true L.A. hyphenates: rock star–movie producer–boxer–tattoo artists. The last particularly

---

[*] Those who have either slept with the staff or are going to sleep with the staff.

significant, since whenever they show up packing ink, tables 10, 11, and 12 become impromptu tattooing stations. The soothing sound of buzzing needles has become something of a post-service ritual—to me, the perfect blend of dangerous and ridiculous. The best piece ever was a headless chipmunk, but that is a story best told by someone else.

Chatter fills the air as we seamlessly fall into our close-out roles.

Rebekka as she wipes down tables: "That table of TV writers left us a huge tip. We're gonna be flush!"

Kimmy as she closes out checks: "Where are my cigarettes?! That first date at table 5 went *too well*. That man's hands kept disappearing under her skirt!"

Devon singing while cleaning her well: "Ohhhh I wanna dance with somebody. I wanna feel the heat with somebody . . ."

Richie as he polishes bottles: "Can I pour some of the Van Winkle, Eric?"

Me as I restock the cooler: "You earned it."

Everyone claps and yells and whistles for Richie.

Ben as he sets up his needles: "Okay, so I have an order for a thunderbolt and a dead cat face. What else?"

Zach to his brother Ben: "You know I'm not good at tattooing, right?"

Office Privilege #1: "Do you guys have tattoos every night?"

Office Privilege #2: "Can I get one too?"

Jeremy: "Don't forget about me!"

Me to the brothers: "Forget about him."

Office Privilege #2 lies down on top of table 10 and pulls off her jeans. "Right . . . there." She points to her exposed left ass cheek.

"One blue rose, coming up," says Ben.

"Why blue?" I ask.

"'Cause of the poem," she says. "Roses are red, violets are blue . . ."

"But you want a rose, right?" Ben confirms, needle quivering in his hand.

"Yeah," she smiles. "A blue rose."

This is one of those moments where I'm not sure the best decisions are being made, but if this woman wakes up tomorrow morning with a blue rose on her ass cheek and a story to tell, isn't that what life is all about?

SINGING. TATTOOING. SMOKING. DRINKING. LAUGHING. Cleaning. We put Jeremy in a cab. Wave goodnight to Carlos with his bag of empties he'll trade in for taco money, meaning what he makes from recycling will go directly to tacos. And when there are no more body parts to adorn, no more cigarettes to smoke, and the sun is threatening to rise, we drag ourselves up and out of the bar. Outside, we say our goodbyes in the fetid morning air, new ink glistening under the sheen of A+D and moonlight, and just as I'm about to turn around and lock the front door, I notice a regular on the street listing as far as one can list without falling down.

A drunk customer is your responsibility. Seems logical, right? Except we don't always take on the responsibility—we've clocked out, have a booty call waiting, or feel fed up because they're always getting too fucked up to function . . .

"Hey man," I call out. "You all right?"

The regular bobs his head.

"You want company? You need help getting home?"

He shrugs.

I sigh. Turn around and lock up, then turn back to him and say, "Let's go!" more energetically than I feel. Thankfully, he lives only a few blocks away, but have you ever walked a drunk person home? It's like walking with a little kid—every blinking light a wonder, every crack on the sidewalk filled with fairies, and don't even try to pass a dog without getting down on all fours to pet it.

"Don't walk like a bitch!" a homeless woman on a bicycle, this time wearing pants, screams at us crossing Main. My regular finds this hilarious, and we have to stop in the middle of the street because he can't stop laughing.

Out of the shadows of a dark doorway, Ricky the Pirate emerges from his evening crash pad. Ricky's been a fixture in downtown since we opened and is always keeping an eye out for people when they leave, making sure they don't get mugged and walking ladies to their cars two blocks over on Los Angeles. He's no saint, but his grift is dressing as a pirate and conning tourists into getting their photos taken with him for a buck, which is pretty tame as far as grifting goes.

"Hey Eric, hey, hey need help?" Eric nods to my wobbly companion.

"I think I got it," I tell him. "Thanks," I say, and watch him do his herky-jerky walk down the street to places only he knows.

AT MY REGULAR'S FRONT DOOR, he digs around for his keys: Back left pocket. Back right pocket. Front left pocket. Front right pocket. Back left pocket. Back right pocket. Front left pocket. Front left pocket!

Alcohol really does bring out the best and worst in people.

"What floor are you on?" I ask as I hold open the elevator door.

"Succhth," he says.

"What?"

"Succhth!" he yells, like I'm hard of hearing, and presses "7." He checks himself out in the mirrors that line the walls and seems pleased with his appearance. When the doors open on his floor, he lurches down the hall toward his front door, unlocks it, walks in, and sits down on the couch. I get him a giant glass of water and watch as he drinks it down. Pour one more and set it next to him and say I have to go.

"Ahhhh," he smiles, and passes out.

BACK OUTSIDE, THE SKY HAS turned an orangey red, blanketing downtown in a Raymond Chandler mist. I wend my way back toward my building, along the shadowy streets—streetlights either flickering or dead. Walking my regular home and seeing the inside of his apartment reminded me of how people come in and out of our lives in uniquely intimate ways. The barista I see every day and notice when he cuts his hair, when he's gloomy and when he looks like he's in love. The taxi driver I listened to a horrible piece of news with on a long ride to the airport and ended up in a conversation deeper than I have with most friends. The trysts I've had with women I didn't quite connect with beyond that night. Even so, when we see each other now, it's part of a shared history that connects us in other, ineffable ways. And there was that summer night in July 1995, when I was a road-worn, weary, post-tripping nineteen-year-old, backpacking around the Southwest, sitting in a plastic Greyhound bus depot seat. My body was encrusted in a thin layer of white gypsum salt from a primal mushroom trip in White Sands, New Mexico, just outside Alamogordo. A thin fifty-eight-year-old woman with very tan skin and a flowery scarf over her head held something out to me.

"Would you like one?" she asked.

It took a moment for my eyes to focus on the cinnamon-colored cracker. A graham cracker! I took it and helped the woman load her bag into the belly of the bus and then gave her a hand getting on board. Her name was Jean Ingold and she, like me, was headed to Salt Lake City.

We took seats next to each other as if we were old companions. At our first rest stop, I bought us snacks and a bottle of water so she wouldn't have to move. She battled a form of Crohn's that made her weaker on certain days and stronger on others.

This was one of her weaker.

We rode through the night discussing life in the way you do when you meet a stranger on a bus. I was reading *The Alchemist*; she said she spent much of her time in the Alamogordo library stacks, close to her trailer. I showed her some postcards from the Grand Canyon, where I'd hiked a week earlier. She was impressed: she'd only been to the South Rim and looked in. I told her I'd studied acting in school, and she showed me a worn photo of her daughter who she thought was in London acting, but didn't really know. When a couple of rough, just-released prison inmates boarded at one of our stops, I kept an eye out for her and our bags.

When I awoke at dawn, my head was resting on her shoulder.

"Write me sometime if you feel like talking," she scribbled her address on a postcard in Salt Lake City. "And thank you for your help."

SEVENTEEN YEARS LATER, COURTNEY AND I were driving back to L.A. from Houston in her father's silver Chrysler 300. Fred had agreed to move to L.A. to stay with us and get sober on the stipulation we bring him his prized car. When we blew through El Paso, I looked at the map and realized we were a mere sixty-seven miles from Alamogordo, where Jean lived. Ever since that day we met, we'd been exchanging letters. About a half dozen annually. A minimalist at heart, Jean had no computer or phone and typed her missives on an old typewriter, signing them, "Always very warm wishes—Jean Ingold."

I'd written to Jean when relatives passed, when I had troubles in my twenties, and as an adult, I'd shared some of my career successes. She wrote about books she was reading, work she did in the community, about her past in a religious cult and her estranged daughter.

When Courtney and I ended up at a little trailer on Vermont Avenue with her address on it, no one was home. There was no mailbox, so I left a note on the door, but knowing Jean walked everywhere, we did a slow crawl around the streets—the Greyhound bus depot, the library, the Waffle & Pancake Shoppe. No Jean. We gassed up and decided to do one more pass by her trailer, and there she was. The reunion was awkward initially, like a first date. Jean had grown into a hunchback from osteoporosis and at first was self-conscious about it. When her nerves calmed, we sat in her one-room home and she shared newspaper clippings, photos, and letters she'd received from other pen pals acquired over the years—a significant network of friends across the country.

When it was time to leave, I asked if there was anything we could do for her. She needed to refill her water jugs from the water dispenser at the gas station down the street. Twenty minutes later, with two full five-gallon water jugs in the back seat of the Chrysler, we returned.

"How do you fill these up on your own?" I asked.

"I pull them one at a time with my yard cart," she said.

The image of this, the hardness of her life, filled me with a deep, terrible sorrow. I wanted to make things easier for her, but had done all I could. And she didn't seem unhappy. She seemed, in person, like she did in her letters—someone doing the best with what she'd been given and the choices she'd made, happy to connect with other humans doing the same.

BACK AT MY BUILDING, I begin my *Goodfellas* walk home through the garage, into the service elevator, and down the hallway to my loft on the seventh floor, which I forever after refer to as "Succhth!"

I shower. Brush my teeth. Drink a glass of water and bring a full one into the bedroom. Put my phone on vibrate and crawl into bed. Take a hit off the vape my brother gave me and lie down. In the dark behind my eyelids, an insistent ticker tape spools a gobbledygook close-out report of the night's goings-on: Everything that happened at The Varnish was amazing. Nothing went wrong. Peter Sellers loved his negroni.

I know I'm dreaming.

AFTERWORD

A bar without people is a room waiting, in suspended animation, for something to happen, for secrets to spill and liquor to swill. Only when it is filled with bartenders, servers, barbacks, and hosts; musicians, regulars, lovers, and neighbors; industry denizens and husbands and wives, does a bar come alive.

In celebration of those who have lit up The Varnish for the past ten years, this space is for them—for their stories to be told in their own words, in the way they remember it going down.

JEREMY KASTEN, VARNISH REGULAR, 2009–2016

Gather round, kids—that's right, onto the bed. All three of you. Who wants to hear a story? Daddy's on his third whiskey and ready to tell a tale! Oh, no—I told you all about Mommy and Daddy's first date *last* week. And we told the story of Mommy and Daddy's wedding just a few days ago. How about the story of Mommy and Daddy's *first* first date? That's right. There was a first date Mommy and I almost never talk about. Do you know why it's a good story? That's right! It was a disaster.

Does anyone remember where Mommy and Daddy lived before we became farmers and moved to the countryside? That's right.

Down. Town. Los. Angeles. And downtown wasn't like a food court at the big, big mall like it is now. It was a crazy, exciting place. And a dangerous place. And it was filthy. Boy, was it filthy. You've never smelled such smells. Pee-ew! Exactly!

Now, in those days, Daddy did lots of silly things because of pain-go-bye-bye juice. And because of cocaine. Everyone in downtown did! It was so much fun, and it was so crazy and chatty! Well, back then there were only a few places Daddy could leave his loft to be naughty and fancy—one of them was a magical place where you were expected to be *both* near blackout drunk and charming. Which was Daddy's specialty back then! It was called—that's right!—*The Varnish*. And do you know how you got into this bar? You vanished *into* The Varnish! Through a *secret door* . . . in the *back of another* bar! I know! Grown-ups *are* crazy!

And when you vanished behind that door, there was a tiny room, and in that room were the tiniest, strongest drinks in tiny, glistening glasses, and in that room were hardly *any* bridge-and-tunnel people in sneakers. Which was *very* unusual for Los Angeles.

Well, as you know, when I first met your mommy, she was a server at a fancy restaurant and it was brunch and Daddy had been up all night. Now. When your daddy saw your mommy, she was poured into her crisp shirt and tiny slacks, as if Fay Wray got poured into a slip. Mommy's Louise Brooks bob blinded in moviestar platinum. And broken-gravestone-chipped red paint on her nails. Man, was Mommy hot. And sassy. And what did Daddy do? That's right! Daddy *did* fall in *love* and asked her over and over to go on dates. The more Mommy said she was too busy, the more Daddy brought dates *into* Mommy's restaurant. And do you know what? It finally worked. Mommy agreed to meet Daddy after her work in the magical bar with the secret door.

Well. You can imagine the look on Daddy's face when Mommy

floated in wearing a party dress, lots of pearls, and long opera gloves. And with her, the handsomest, most charming, *skinniest* male-model-looking fellow you've ever seen. Mommy told Daddy he was her "friend" visiting from out of town. So do you know what Daddy did? That's right! He got a really strong drink from his friend the server whose name was Kimmy. And then he asked her for *another* really strong drink! And then he asked for *another*. And soon Daddy didn't care what happened! And *oh my, it's time for you two to leave? It's so early! Nearly closing? You're kidding! Well, by all means. This has been lovely. I think I'll stay for just one or six more, but I'll walk you two out.*

The next thing Daddy remembers is standing in the brisk November air outside the bar that hid the magical Varnish bar, with the streets streaked with gritty L.A. rain and pee-pee. And suddenly Daddy didn't feel so funny or loud. He felt small. And lonely. Daddy remembers thinking, *Well. That's where that ends.*

And do you know what he did?

Daddy went back into the bar. Through the door in the back where he vanished into the tiny, magical room. And asked his friend Kimmy for another really strong drink. And no doubt Kimmy told Daddy a lie she'd told lots of customers before.

A disaster? It looked to me as though she liked you a lot. Oh, come on! She was laughing the whole time. Call her tomorrow. Do you want another one?

And that's the story of Mommy and Daddy's *first* first date.

MATTY EGGLESTON, VARNISH BARTENDER, 2009

Comparable to foggy recollections of a strong night out, my memories of The Varnish are both distant and distinct, but not a workday goes by that isn't influenced by my time there. Like that oft-quoted aphorism about folks not remembering what was said or done but remembering how they *felt,* being asked to join the opening team was a benchmark career moment.

The Varnish was so much about bartending that I loved at the time: the precision of the room, the backbar, and the stations. The cocktail specs. The crew. The first jangly notes of "Straight, No Chaser" as the door was unlocked. Slipping a None but the Brave or a Champs-Élysées into the mix for Bartender's Choice. A garishly big knot for the necktie. The healthy level of nerdery. Stuff that's everywhere now. Or come and gone already. The whole wheel is turning around again on what is de rigueur, but at The Varnish it was fresh and deeply rooted, and the thing we made was tuned as tightly as possible to ensure a roomful of people got loose on cocktails.

And it was fun.

REBEKKA JOHNSON, VARNISH BARTENDER/ SERVER/HOST, 2009–2013

I was a loser working at a small chain-style brewery that had fourteen different kinds of salad dressing and an always creamy soup of the day. I had moved to Los Angeles from New York City three months before and was trying to navigate Hollywood while slinging beers and fried chicken salads. My best friend and comedy partner Kimmy Gatewood invited me to grab a cocktail at her new job. So on a day off, still smelling like ranch and strawberry blonde beer, my husband and I went to The Varnish for the first time. There is something magical about candlelight and crystal lamps; I felt transported back in time to a fictional pre-Prohibition bar—bartenders wearing suspenders or pencil skirts with red lips. I knew I wanted to be a part of this drunk old-timey family.

Kimmy introduced me to Eric, and I immediately confessed I

had to work there. He took my number and called me the next day to give me a job. He said he liked my 917 number and New York attitude.

I had to battle to keep that attitude in check when dealing with impatient people at the door pretending to be investors and rude men who didn't like that we served daiquiris in a "girly" coupe glass. But mostly, I had fun. Probably too much fun. Sometimes the fun made it hard to count out the money at 2 a.m. Even though I had come to L.A. for a life in comedy, The Varnish was never just a day job (or night job). It was a cool place to work. A place I was proud to invite my manager and agents to. A place I went to immediately after performing on *Conan*, in full hair and makeup, to work behind the bar. A place that invited me and Kimmy to sing our original 1940s-style comedy songs one New Year's while taking a break from serving the best cocktails in town. Ryan Gosling used to frequent The Varnish and became a fan of our comedy group, The Apple Sisters. He went on to ask us to open for his band and produce our album. It was a magical time.

Four and a half years later, my comedy career was starting to take off. I was the producer and director of *Speakeasy* with Paul F. Tompkins. I was acting and writing but still cocktailing one day a week. I finally quit when I was two months pregnant with my son and would nearly throw up from the sight of whiskey. And there was so much whiskey.

I've had so many full-circle moments connected to The Varnish: I waited on Alison Brie and now I wrestle her on *GLOW*. I waited on Sandra Oh on a slow Halloween night (dressed as a Dark 'n' Stormy), and this year at the SAG Awards we were both nominated. I chatted with her on the bathroom line. When I mentioned that I waited on her so many years back, she was pumped. She loved The Varnish and loved that I was there then, and here now.

It's been five years and I still feel at home in the bar, and Kimmy and I still brag about working there, offer to make executives drinks at pitch meetings.

Is that weird? Who cares. I'm cool now. Don't you think I'm cool?

(BTW: I still smell like ranch. Ranch and whiskey.)

SIMON FORD, FOUNDER OF THE 86 CO. AND FORDS GIN

For me, The Varnish was the bar Los Angeles was waiting for when it finally opened its doors in 2009. In my opinion, The Varnish brought a slice of New York drinking class to Los Angeles at a time when it needed it most, and it had a dream team to get it going—Sasha Petraske, his friend Eric Alperin, and Cedd Moses, a downtown L.A. visionary with a great eye for design and a passion for the types of bars and drinks Eric and Sasha were known for. As soon as I heard of its opening, I was there. A lot. It was my home away from home whenever I visited Los Angeles, the kind of place I knew I would get a perfect drink and was guaranteed to avoid that classic L.A. velvet rope bullshit and pretentiousness that was such a big part of the city at that time (although I did secretly enjoy those places as a guilty pleasure every now and again). I would find myself hanging around at the end of the night, hoping to grab late-night tacos somewhere with them after they had closed. It was like the bar on TV's *Cheers* for me, where the bartenders knew my name and what I liked to drink.

Knowing all of this, it should come as no surprise that I have many great memories and associations with The Varnish, and it is also a bar that had a great influence on me. Every time I stepped behind the bar I was taken aback at how well designed it was, and how intuitive. Everything was where you would expect it to be, and even though I didn't ever work there, I found everything I needed to make a drink quickly and easily. So much thought had gone into every tiny detail. It isn't a flashy bar though—it's purposeful, like a classic vintage car everyone wants because no one ever improved on its design. The Varnish was an instant classic. It was also a big influence on me when I was creating my gin. Sasha had already helped me create the recipe, but when I was working on the bottle's design and labels, it was a day spent with Varnish bartenders that set things in stone: the ridges on the neck of the bottle came from Eric showing me his "London pour"—it became evident that we needed to make the shoulders of the bottle more ergonomic so it would be comfortable to hold when pouring. The Varnish's attention to detail was passed on to me, and to this very day people thank me for the bottle's design.

MIKE BITTON, FRIEND OF TWENTY-PLUS YEARS, BARTENDER AT THE SLIPPER CLUTCH

I remember when Eric was working at Osteria Mozza, I must have been going there anywhere from one to three times a week, and you know I had the mad hookup. So my first thought when I realized Eric was leaving Mozza to open The Varnish was, *Well, there goes my fuckin' hookup.*

Obviously, I'm only kidding. Sort of. The fact that my homeboy was opening his own spot really made me happy and proud. And who knew that this little spot in the back of Cole's would actually change my life in the most profound and positive way, 'cause seven years later, Eric, Richie, and the 213 family opened a bar called The Slipper Clutch. Not only did they give me a job, they gave me a whole new family and social life. Aside from opening a dope fuckin' place with such drive and a fun crowd, they gave me a family—Christina Ray, who I work with, has become my sister, and Josh, Kevin, John, Armando, Gustavo, and Dustin.

I make a good living there now, but more important, having this crew in my life makes me feel extremely happy, makes me excited to go to work, which I never had in my life. And I also found that I'm good at what I do—I'm forty-eight years old, and this is my first gig in the service industry, and because of The Varnish (where it all started), I've found my calling in life.

Oh, and I am also an ex–heroin addict and Eric knew me then, so the fact that he can see me at my lowest and now see me at my highest is amazing. And remember I said I was upset about losing the hookup at Mozza? Well, now that I'm in the industry, I'm pretty fuckin' hooked up everywhere.

RICHARD BOCCATO, PROPRIETOR, DUTCH KILLS,
FRESH KILLS, HUNDREDWEIGHT ICE,
BAR CLACSON, AND THE SLIPPER CLUTCH

On Saturday, April 7, 2012, me and Eric Alperin rode our motorcycles from Baja California up to Los Angeles. We parked the bikes in the garage behind The Varnish and settled in at the bar for the night. A kid who used to work for me was tagging along. About an hour or so before closing time, the Shields brothers showed up—Zach and Ben Shields are tough as nails. Both accomplished and uncompromising artists. They were accompanied by an attractive blond woman to whom Zach was married at the time. And they'd brought their tattoo guns.

In the storage room behind the bar, while service was still in swing, Eric got a tattoo that resembles a death mask of a cat's face. The kid who used to work for me got a tattoo; I forget what it was. When it was my turn, I relaxed my arm on a prep table above a chest freezer and watched as Ben carefully scratched away. He drew the face of Fritz the Cat, from the Ralph Bakshi movie, on the underside of my left forearm, just south of my elbow. His whiskers are delicate and lifelike. He wears a mischievous grin and a bow-tie. Niki came back to let us know that the last-call patrons had left, and we all spilled out into the bar.

The show was over. The lights were up. No music. The guy working the door locked up and brought his friend into the bar to get a tattoo on his neck as me, Eric, and the kid who used to work for me drank Tecates and mezcal. Niki was counting the register with a lit cigarette in her mouth, wincing through the smoke as the paper money slowly slid through her fingers from one hand to the other. The young lady to whom The Varnish's GM, Max, was married at the time was reclining on a table with her pants down as Zach outlined a chipmunk high up on her thigh. She was giggling. A tobacco haze was hanging in the air. Needles were buzzing. I remember seeing Zach's wife rummaging through her purse. Soon she was really going at it, looking for something that should've been there. She started saying something about how the door guy's friend had taken her wallet, and Zach's tattoo gun stopped buzzing, mid-chipmunk.

The door guy's friend wasn't in the room. He wasn't in the bath-room where Charles Bukowski and Mickey Cohen once pissed. Zach and Ben both reached into their respective bags, and each pulled out something similar. I didn't see what it was. Then they left the bar, and Eric followed, shouting on his way out, "Richie, you're in charge!"

I swallowed some Tecate backwash, stood up, and told every-one, "Let's clean this bar!" Because that's what you do when after-hours is over. You clean the fucking bar.

A few weeks later, back in New York, I went to a movie on the Upper West Side.

After it ended, I stayed in my seat to watch the credits roll. Ben Shields's name made its way up the screen, under the title "Tattoo Designer." I looked down at Fritz on my arm and thought, *There's a young lady out there in Los Angeles tonight with a headless chipmunk hiding just above her panty line.*

JAMIE ELMAN, VARNISH PIANIST, 2009–PRESENT

It's not hyperbole to say that The Varnish changed my life. I was thirty-two and had been working in Hollywood as an actor since I was twenty-three. I had never had a professional music gig in my life, although I had played piano and sang on TV in a couple of shows over the years. Lifelong "hobbies" of mine, my passion, sure, but not my career.

Most of my piano playing in L.A. was with a great group of guys that I jammed with on Saturdays in a pool house out in North Hollywood. We called ourselves PoolHouse—appropriate, since we never played anywhere else.

The guy who invited me to these stoney sessions was the drummer, Johnny Sneed. Like me, Johnny wasn't a "professional" musician but rather someone who just loved to play. So when I got a call from someone named Eric Alperin, telling me he got my number from Johnny, I was immediately nervous.

"Did Johnny tell you about my playing? Cuz we really just jam and I'm sort of a blues guy, not really a professional. It would probably be best if you heard me play first? Like an audition? . . ." I stammered.

"Sure man, if you wanna swing by here, come on down," he replied.

That was my first time in downtown in the eight years I'd lived in L.A. It was grand and dirty and felt like a different planet, even though I lived in Silver Lake, just fifteen minutes away.

Inside The Varnish, the bar was in pieces. Eric was dragging benches around the room. The lighting fixtures were not in yet, and in the middle of the room was a battered old upright. Eric was friendly but distracted. I played a few songs I thought would be appropriate as he took phone calls and moved stuff around.

After about fifteen minutes, I said, "Something like that?"

"Sure, sounds good," Eric replied.

February 24, 2009. Opening night of the bar. I was overwhelmed with excitement and nausea all day. When I walked in, I was expecting a small group. It was packed. I was freaking out. I felt like a fraud. I was wearing a shirt/tie/vest combo, with a cheesy fedora—*I may never be asked back*, I thought, *but at least I look the part.*

The bar was gorgeous. It was so cool. I tried not to attract much attention. Prayed that no one would talk to me or make any requests. Was comforted that it was too loud for anyone to notice I repeated the same songs all night. Driving home, exhausted and relieved that it was over, I assumed I wouldn't be asked back, but when I checked in with Eric, he told me to start that Monday.

"Really? Cuz, like, it's my first gig and I'm just not sure—"

"It's gonna be great man," I remember him saying. "We'll figure it out together."

On my two hundredth Monday, I invited all the people who had ever sat in with me to come celebrate my self-proclaimed milestone. About fifteen different players swapped in and out that night—one of my favorite moments in my nearly twenty years in L.A.

Eric told me recently that I "have a lifetime residency."

I intend to take him up on that.

KIMMY GATEWOOD, VARNISH SERVER/HOST, 2009–2011

My job interview with Eric was at the bar counter at Cole's French Dip. I ordered a grilled cheese and soup. Eric didn't seem like the bar owner of what would start the cocktail revolution in L.A. He was just a dude with unkempt hair and a lot of passion. We talked comedy, writing, and dreams of making movies. Next thing you know, I was training as a host and waitress, under the care of Deb Stoll, who I knew from New York.

I found out I was the first "external hire" who wasn't family or "cocktail family." I felt so proud of myself for that but had no idea how important and special this family was going to be.

In the early days, we did everything by hand, from making the cocktails (always) to chopping ice to writing the drink orders to counting the cash. We had to taste and describe every single drink that was in the phone-book-size book of classic cocktails. And then as new bartenders would come in and create their own cocktails, we had to remember all of theirs. At some point in the night, I would lose track of how many drinks I had served and how many I had marked off and how many had been paid for, so I'd just guess around ten drinks. Sometimes I'd overcharge and undercharge customers by accident, but no one seemed to care. Poor Eric would be there *every night*, reining us in, tasting the drinks and making us laugh. Or he would be stuck in the smelly, smelly back room, doing paperwork for his employees, who kept very bad books. I learned: *Don't put your stuff in the back room if you have to be somewhere*—if the back door was locked, someone was either doing drugs, making out, or getting oral. I never partook in any of those shenanigans, and looking back . . . I'm still okay I didn't. The back room smelled like French dip, lemons, and trash!

Some Yelp reviewer gave me a scathing review, accidentally calling me Kippy. It was truly a crowning moment, because you know what? They were fucking wrong.

My favorite thing ever was naming drinks. Alex Day came up with a drink and asked me what to call it. I told him "the High Five." He asked why. I told him because of the garnish—a high five. One day I'll win an Oscar, but naming the High Five will still top it. (Unless the Academy is reading this, in which case I retract that last sentence.)

It's weird to look back and realize I was in the L.A. cocktail revolution when it was happening. We were having so much fun at work. We would drink too much, dance around, sit with the customers, yammer on for hours about cocktails, and pretend we were showing discipline when we wouldn't serve cocktails after 1:30 a.m., but do a hundred shots, open beers, and smoke cigarettes in the back.

I got to hang out with famous chefs and celebrities who wanted to try the best cocktails in the city. I met and served Jonathan Gold, who invited me to sing at an event, and he put The Apple Sisters on a spread in *LA Weekly*. We sang songs on the piano with Ryan Gosling and Olivia Wilde. We threw back beers with Ludo and his entire kitchen staff after they opened up their first L.A. restaurant.

The two years that I was there, we developed a tight-knit family that fed the nightlife in Los Angeles. I love The Varnish and my Varnish fam.

ALEX DAY, VARNISH BARTENDER, 2010–2011,
CO-OWNER, PROPRIETORS LLC AND DEATH & CO.

It started like any other shift at The Varnish: I sauntered in around 4 p.m., iced coffee half empty and the last lingering bits of a hangover on the tip of my brain. I recall beginning the normal routine: ice always came first. Back in those days, we cut from big blocks frozen overnight in massive chest freezers that filled the bar's tiny office from wall to wall. I'd listen to LCD Soundsystem or the Shins or whatever else I was into back then. The music didn't really matter, but it was critical for keeping pace and moving through the task, my mind wandering to thoughts of life and purpose and a place in the world as my ice pick precisely skewered frozen water in rhythm.

All those beautiful pictures of cocktails served on a perfect, clear cube? Yeah, there's a lot of working shit out in the process to get there. Namaste.

Once we finished with the ice, the lights were dimmed, the candles were lit, and we were off. The first few hours were typical enough. Rounds of old-fashioneds and daiquiris, Queens Park Swizzles and negronis, and always a few early-evening martinis for the classy crowd. Around the tipping point, in walked a couple flanked by friends—the guy young and brunette and extremely handsome in a perfectly fitted suit; the woman petite, with a flowing dress, a determination in her step. We all recognized them immediately.

I took them in for a moment, then realized Devon was waiting impatiently for her drinks on the other side of the service pass, snapped out of it, and got back to work.

The Varnish stops for no one.

At last call, everyone shuffles out the door, but the couple stays. "You guys want anything else?" someone asks. "Sure!" One last round for Drew Barrymore and Justin Long, who, I soon learn, came to us directly from a movie premiere that had bombed fantastically—their big onscreen co-starring role together. Despite this, they're having a blast, in the throes of a romance, our drinks helping alleviate their disappointment.

The couple abandons the solitude of their table to join the staff. As was our custom in those days, a few shots and some beers sped up cleaning and paperwork (maturity has proven this less than correct), but it was also a time to air the night's stresses, let out the chaos of the night, reflect together, get a little tipsy, and come down off the high of making other people happy. They joined us, polite and engaged, chatter and banter in equal measure. *Why are these drinks so good? How long did it take to learn this?* We slough it off—*It's really easy, just a few things to memorize, attention to detail, great ice, no big deal.*

Fueled by laughter and mezcal, I invite Drew behind the bar. Someone grabs ice from the office and we get to it. *Here's how you hold a jigger. Here's what it means to be accurate. Here's how we shake. No, no, shake harder—it's block ice. Harder!*

And then Justin takes his turn. More cocktails are made. Things get fuzzy. I make a joke that Drew is the only person around smaller than I am, and before I know it she's climbing around my torso. It is one of the strangest and funniest things anyone has done in my life, the type of innocent playfulness that speaks to a person's true self and heart. Eventually the laugher slows and sharper minds put a halt to the booze, we bid farewell, and the evening ends, as it always must.

That night wasn't memorable because two famous people stayed up late with a few rowdy bartenders and servers. The night certainly wasn't memorable because we stuck around and drank until an hour we shouldn't even admit. That happened all the time. That night in 2010 when Drew and Justin joined us into the quiet hours was one of the most memorable Varnish shifts because it set in stone in my mind the immense power a bar has to bring people together, no matter who they are, away from worry, and set them free, if only for a few precious hours.

ERIC NEEDLEMAN, PARTNER, 213 HOSPITALITY

I can't recall the first time I stepped into The Varnish for a drink, but it was surely sometime in its first year, 2009. I wasn't a cocktail neophyte, but I very much believed the drink was all about the ingredients and the technique with which it was made. In fact, The Varnish did use only high-quality ingredients and a very curated spirit selection, many of which I hadn't seen elsewhere, to accommodate its small backbar. The tenders there were quite proficient, measuring amounts accurately and using proper technique when stirring and shaking. The drinks were delicious, and I chalked it up to those two factors: good ingredients and good technique.

Then I began to notice others in Los Angeles making "proper" cocktails, no doubt at least partly inspired by visits to The Varnish. It was at that point I realized that anyone could buy those ingredients, anyone could refer to books for classic recipes, and anyone could learn the basic techniques for making a good cocktail. And the drinks were good . . . but not quite as good as the same ones at The Varnish.

As I spent more time there and helped open Half Step, their sister bar in Austin, Texas, I began to learn the subtle details. First, technique is more than just shaking and stirring properly. I learned that the order in which a round of drinks is made matters. I learned that making your own ice in large, pure blocks provides a colder drink with much less dilution. And I learned that it's important how glassware is handled so body heat doesn't warm up that drink.

Guests don't notice these things, but they sure contribute to why that drink tastes so damn good. More than that, though, it's the experience of The Varnish that makes the difference. It's the host who greets you upon entry, acknowledging your presence and immediately providing a sense of welcome. The warm glow of the lights throughout, creating an intimate and comforting environment. The standing bar with extra-wide arm cushions, the custom insulated garnish trays that maintain fresh products within arm's reach of the tenders, and the soft leather cushions you sink into when sliding into a booth.

These are all acts of intention. Every last one of them, in the spirit of presenting the best version of a cocktail and creating the best possible environment in which to enjoy it.

So are the drinks at The Varnish made with great ingredients and proper technique? Absolutely.

Is that why the drinks always taste better there? Not one bit.

KATE GRUTMAN, FRIEND AND FORMER CO-WORKER

In 2012, I was at Tales of the Cocktail in New Orleans as a rep for Anchor Distilling Company under Southern Wine & Spirits, and The Varnish won Best American Cocktail Bar, and E, A, J, and I got so turnt and on the shoulder (plastered/smashed) we ended up singing Journey on top of the bar at Erin Rose, then stripped down to our drawers and ran through the kitchen and into their meat locker. Security escorted us out and we all ran in and did it again. Then I fell off E's bicycle handlebars on the way to the "house," where C and a bunch of people I don't remember all partied until G lost his grip on a bottle of Ocho Reposado and gave me a minor concussion. Everyone kept me awake by throwing me into the clap/gonorrhea/hepatitis pool on top of D and A, and of course K, and then I woke up, fully clothed and chlorinated, on top of a plantation-style canopy bed, and when I went to look for E or C, I found two different bedrooms covered in vomit but otherwise empty and one bedroom with at least nineteen people passed out in a dog pile on one bed, and that's how you end up coming home from an industry event with strep throat and a black eye, still on retainer with Anchor and repping a rum called Pink Pigeon that came bottled and branded with a pink cock ring.

ERIC THORNE, GOOD FRIEND AND MASTER CARPENTER

We struck a bargain
    I never paid for a drink
    Would work for cocktails

CARLOS LOPEZ-FLORES, VARNISH BARBACK
FROM DAY ONE TO TODAY

Cuando mi hermano Héctor me comentó que su amigo Eric iba abrir un bar, me dijo que está buscando personas para trabajar. Yo le conteste a mi hermano que tenía un poco de inseguridad por que nunca había trabajado de barback y no sabia inglés. Pero mi hermano me dio seguridad, me dijo que Eric le gusta la gente que trabaja rápido, atenta, y eficaz. Yo me sentí confiado y pude ir a la primera semana de entrenamiento.

Mi primera reacción fue intimidado, porque había mucha gente familiarizados con el trabajo. Y a pesar de que la gente me enseñaron bien, todavia me sentia fuera de mi area. Al final del dia le comente a mi hermano sobre del apertura del bar, no tenía fe en el negocio porque está muy escondido y la arena estaba muy sólida. Rápidamente me corrigió mi hermano: "Estamos trabajando para el mejor bartender."

Después me llamaron para darme solo un dia de trabajo. Pero ese dia fue terrible para mí, no entendía el sistema y tuve que hablarle a mi hermano para que me ayudara a cerrar. Pero por suerte la siguiente semana estuvo mejor, me aprendí todo muy rápido de la semana pasada. A mitad de turno me preguntó Eric si podía venir a trabajar el viernes, quiero pensar que el vio mi potencial para trabajar. Ese viernes llegó y me sorprendió ver que el schedule decía que iba a trabajar seis días.

Despues de dias, semanas, meses, e incluso años me gané el respeto y la confianza de Eric. Al que me a dado un apodo "Papa," me puso mi foto en su pared de honor.

RIP Sasha Petraske

. . . por que uno de los grandes me enseñara a trabajar.

SARI LINDERSMITH, VARNISH BARTENDER/
SERVER/HOST, 2011–2018

In the fall of 2018, I was trying to compose my resignation letter. After seven and a half years, I was leaving The Varnish. In the same moment, I received an email asking if I'd like to make a contribution to this book. After a few seconds of reflection, I burst into tears.

I started at The Varnish in 2011. I was twenty-two years old, with close to zero service industry experience. I was a big fan of the bar as a patron, but my main career focus was fashion and I resented the fact that I had to work nights to supplement my income and intern during the day.

I graduated high school in the height of the recession. It was a time when guidance counselors basically shrugged at you, because every industry seemed to be dying. I attempted more than one career path before finding myself at The Varnish, but it took working there a few years before something clicked. Sasha came to Los Angeles to do a training with us in early 2014, and I was so nervous I had to take a Xanax. We were each instructed to make a martini for him while he watched our technique, tasted it, and tossed it out. Martinis are my absolute favorite drink, which made me even more nervous. When it was my turn, I stepped behind the bar and made my martini while thinking about everything he had told us about agitating the ice and checking the temperature often. One thing he said that stuck out to me the most was: when you sip a martini, the first thing you should experience is very cold water; the other flavors should follow.

When the martini was ready, I strained it into a glass and placed it across the bar in front of him. He took a sip and very casually said that this one was too good to toss. Then he asked me to make him a grapefruit twist. He expressed it over the martini and handed it back to me to taste. It tasted like very cold water, and all of the other flavors followed. This experience made me realize I was no longer just working a night job to afford my low-paying fashion gig. I realized that I loved what I was doing and I was good at it. I realized I had found my people and my path.

Soon after, I left my other job and committed the next several years to The Varnish.

GORDON BELLAVER, VARNISH BARTENDER, 2013–2018, AND PARTNER IN PENNY POUND ICE

In my time, I have seen The Varnish's napkins, which are in fact dental napkins, worn as bibs. I've seen people put them on their face and cut out holes and make masks. I've seen business proposals materialize on napkins. I've watched people unfold and refold and flip their napkins expecting to find some magic treasure hidden beneath. I've had someone ask me if the menu is inside the napkin. (It's not.)

Have napkins been lit on fire? You better believe it.

As for me, I've used napkins to cover bleeds and clean lipstick off the rims of glasses. I've used napkins to swat annoying customers. Napkins have been bracelets, table wedges, and moist towelettes. Need to clean a table quickly? Use the condensation from the glasses to mop up the sticky spots.

Napkins can have their own visual cues. Placing them atop a drink means *I'll be back*. In the case of The Varnish, leave a pile of napkins stacked in the corner of the table to inform the next employee, *Hey, this needs a wipe-down from a real towel*.

Napkins are presented when a table needs to be seated. Napkins are where the drinks go. Napkins are the best way to give someone a telephone number. Napkins mark the passage of time. I had a relationship with napkins for almost five years at The

Varnish as a bartender/server/host, before transitioning into my current position as a partner at Penny Pound Ice. When the time came for me to depart, I took with me countless memories but only one napkin, for that napkin gave me my first marriage. He didn't have the gall to give it to me himself as he stumbled out of the bar. At the time it seemed rude of him to leave without saying goodbye, but on his napkin he'd left a note, "Thanks for the shots," and a number.

It was a full week before I sent him a message, not out of some power move but simply because I had placed the napkin in my pants pocket and didn't wear those pants until the next Saturday. A few weeks passed until we actually spent time together, and months before we were exclusive. Engaged for years and married for years, all while visiting The Varnish on a semi-regular basis. We had highs and lows there, and I can recall moments of twirling the napkin absentmindedly between my fingers like a nervous tic.

When we separated, I looked at that napkin long and hard, debating what to do with it. I used to show it off to friends and family when they visited, a veritable sign of our courtship. I had always wanted to frame it, but there was something in the tactile sensation of touching the napkin that reminded me of him.

I still cling to that napkin, buried somewhere deep in memorabilia from years past, knowing that one day I'll flip it over and start fresh, but not ignore the stains on the other side.

DANIEL ZACHARCZUK, VARNISH BARTENDER/
SERVER/HOST, 2011-2015

There is a romance to arriving at the bar, to greetings and orders, to cutting ice, to holding your glass, to shaking and stirring, to candles, to music, to being in range, and there is a romance to leaving but never being gone.

ANONYMOUS, A FRIEND OF MANY YEARS

I have cut away a tangled parachute, reentered free fall, and deployed my reserve. I have jumped into a glacial crevasse. I have spoken in front of a crowd of thousands and frequently swim in the open ocean with sharks. Nothing in my life has approached the feeling of fear I had while I stood and faced the door to the room of my intervention.

Six months later, I stood behind the bar at The Varnish, off-hours, and rifled through the beer cooler looking for water. The shimmering bottles of liquor on the backbar gave me pause. There had indeed been good times, even magical and fantastic times in The Varnish and in other bars, but as my life with alcohol went on, it became like playing grab bag with a sackful of hand grenades—when one went off, someone I loved was usually hurt more than me.

For most of my life I felt I needed to drink to talk to women, to speak with confidence, to make and keep friends, and to be a man. Alcohol had informed much of my personality from a young age, and until my late thirties I could not conceive of a life without drinking.

Then it happened. At thirty-nine years old—an intervention.

Rehab.

And the next summer, sober, a six-thousand-mile solo motorcycle adventure that brought me all the way back to The Varnish.

So there I was, like an ostrich, head down in the beer cooler. Behind one of the Czech lagers, I grasped the finest bottle of club soda known to man; this man, anyway. I clinked my club soda against Eric's glass of mezcal and we had ourselves a proper toast and I found being at The Varnish sober to be affirming and empowering.

Four years later I am still living sober.

P. J. PESCE, VARNISH REGULAR AND
FILMMAKER, 2010–PRESENT

When I was growing up, I was a huge fan of *The Lion, the Witch and the Wardrobe.* The idea that one could find this little rip in the fabric of the universe and walk into a time out of mind, place out of space, where everything stops and the world is an idealized version of "not here, not now." I'd heard The Varnish was such a place and dragged my partner with me to see if it was really true.

And it was. Every drink was extraordinary, as if the person who made it had taken great pains to make it balanced, subtle, and gorgeous, done with love. They gave me not only a drink but an experience, and it was exactly what I needed. I felt a deep connection to the place, and to whoever had created it. I told my wife that when we redid our kitchen, I must have something like it—that our kitchen must reflect this gorgeousity, this profound obeisance to the glory of alcohol. What proceeded was one of the worst fights of our married life.

Are you out of your fucking mind? We can't make our kitchen into a bar.

We must get a divorce. You obviously have no understanding of who I am on a fundamental level, of what is important to me, and cocktails are deeply and profoundly so.

Somehow, through the haze of multiple incredible cocktails (in spite of or because of), we survived.

A few years passed and I was directing a TV show with a great actor, Anil Kumar. He came to our home to watch the show and of course, I made everyone drinks at my bar. He was impressed.

"What's your favorite bar in L.A.?" he asked. Unhesitatingly I replied, "The Varnish."

"You're kidding! One of my best friends owns that place! We went to acting school together!"

Two weeks later I found myself seated next to a man who would become one of my dearest friends, someone who shares my love of food, drink, friendship, motorcycles, women, movies, and old-school honor.

When my father died and I needed a place to lick my wounds, Eric and The Varnish were there. I drank deep, the alcohol dulling the pain. And when the bill arrived, there was a note on it:

"Pops got drinks tonight."

I consider The Varnish one of the most sacred places on this Earth.

MARC GASWAY, VARNISH MUSICIAN, 2012–PRESENT

When I started to take responsibility for the music in 2015 (I was already calling myself The Varnish's "musical director," at least on my Tinder page, but would never have called myself that in front of Eric), I started bringing in some of my own compositions. Just instrumentals at first, but eventually I knew I could write songs that fit well enough nobody would know I wasn't singing a cover.

I'd go to The Varnish on all my off nights and order a Bartender's Choice, telling whoever was working what I was writing about, and asking for a cocktail to echo the mood I was going for: For "Oarsmen," a song with nautical themes, I'd wind up with a dirty martini, because it's briny. For "Marmalade" I'd sip a Singapore sling, because it's a song about two people screwing on a picnic blanket while drinking Singapore slings. I'd wait for Max to give me the green light to take a booth for myself and sit down and put pen to paper.

Now we've got a record out—*Live at the V* (I'm a bit paranoid that people are going to think that's what people call The Varnish, so let me say now, it's not. Nobody calls it that. I just needed the *V* so I could make a lyric on the record rhyme properly.) The album was recorded in front of a live audience at The Varnish, where all the lyrics were written, and features all the musicians in our Varnish music family.

It's impossible to separate the music from The Varnish's essential connection to my life. Those songs couldn't have been written anywhere else. I've spent more hours playing and creating music at The Varnish than I have anywhere else. The room itself is in the music's DNA.

CHRIS ANGULO, VARNISH REGULAR AND SCREENWRITER, 2009–PRESENT

It was October 2018 when I brought my girlfriend to The Varnish. We had only been going out for a few months and I hadn't brought her around my friends yet, but I wanted to show her The Place, let its classiness leave its scent on me. Not that I needed it, but it couldn't hurt.

She'd never seen anything like it—a super-dope place hidden in the back of another place, dark, old-school, guy was on the piano playing, the host knew me, we got a table right away—I was classier by the second.

When my girl asked how I knew Eric—perfectly normal question, nothing to see here—I tried to put a button on the moment with a cool answer, so I thought for a beat.

What to say . . . I had met Eric eight years before while secretly working on this movie. Long story, but the lead actor had to make the perfect old-fashioned, and Eric was the only guy to teach it, apparently. I went to the lesson, too. Which turned into a drinking session, and soon I was making fun of him for no good reason, a little-brother thing I can't shake even in middle age.

Then I thought about the time Eric watched me die at a birthday party in Catalina. I had only known him a few months and had this strange accidental-on-purpose Hunter S. Thompson deal going. I was on the slide down when Eric and I were alone at the condo. I went into the bathroom, and minutes later there was a crash and a sickening thud. Eric ventured inside to find I had fallen through one thick glass shower door, and the other appeared to have bisected me at the waist. I sat contentedly in the tub, feeling absolutely fine while he took care of me.

So there I was at The Varnish, sitting across the candlelit table from my perfect newish girlfriend, all smiles, her big browns looking at me, her hand in mine, waiting for an answer.

"Eric? I think Zach introduced me or something."

JOHNNY SNEED, VARNISH MUSICIAN, 2009–2017

I am forever grateful for the many Monday nights I've played at The Varnish, a magical place where drinks, conversation, and music stimulate and inspire all who are fortunate enough to be there. Many musicians have joined us over the years—professional and novice alike. Jon Brion playing piano and guitar with us was a definite highlight. Another happened to be when my parents were visiting: Jamie Elman was on piano, Marc Gasway on bass, Claire Wellin on violin, and I was on drums. Claire sees the door open and then leans over to tell us we have to play something especially good, *right now*—Chris Thile, the Grammy Award–winning mandolin player and current host of *A Prairie Home Companion*, was walking in. He sat down at a table near us, mandolin by his side. After he had a drink, we invited him to play with us, which he did to much applause and several flashes from my mother's camera.

The Varnish is that kind of place—where my mom and a Grammy winner can share a drink and a laugh while singing songs together.

LINDA AND DAVE, VARNISH REGULARS, 2010–PRESENT

"Living on borrowed time." This is how we like to describe the interval we inhabit at The Varnish before the inevitable blackout sets in. Once we enter, we accept that only fragments of our experience will come floating back to us in the morning, among the drunken flotsam of random cocktail pics, matchbooks, and hopefully a receipt to confirm we actually paid for everything. "Living on borrowed time" is what Linda announced to Devon as she blessed a batch of shots she thought we needed, and it has stuck. It's what Linda was thinking, no doubt, when she suggested Anthony smell those rubber summer sandals with the special manufactured Strawberry Shortcake doll–like scent, and he, without pause, held one damn shoe aloft and inhaled its perfume. It's what David has been up to when he's been *that guy* and slurred a request for something Chartreusey like a Greenpoint or a Bijou but not exactly either of those.

Navy-strength gimlets, amaro with staff, snaquiris, shots from an ice luge in the shape of a giant penis—we've taken a lot in our borrowed time at The Varnish. But maybe nothing more so than its logo. To commemorate our drunken hours together and our marriage, we decided to get matching cocktail-themed tattoos, and when text tattoos like GIN & on his forearm and TONIC on hers didn't seem cool enough, we ripped the image of the coupe from The Varnish. Proud of this new ink and how each cocktail glass tilted toward the other in a tattoo toast, excited to show friends outside The Varnish's circle, we were nevertheless worried—had we become worse than those obnoxious regulars we sneer at when visiting other bars? Even more terrible, had we become stalkers? Would the people there view us differently? Would the smiles we saw as they chipped ice at the bar take on a more sinister meaning?

And what about Eric? Speaking of stalkers, he had just moved into our apartment building, taking up residence on the same floor. One of us embarked upon elaborate measures, never rolling up his sleeve, turning his arm inward whenever he detected a fashionably professional masculine presence in the hallway of his building.

Perhaps we took ourselves too seriously. Surely we underestimated the pride that comes from being a part of The Varnish. Eric eventually discovered the tattoos, and instead of banishing us from his bar for being creeps, he rewarded us with a big cooler full of Penny Pound ice.

And so, with the exception of one insignificant blip, we have persisted in drinking at The Varnish, comfortable in our roles as two fortysomethings who appear semi-regularly and get hammered and yammer on about borrowed time.

MIKKI KRISTOLA, VARNISH BARTENDER/SERVER/HOST, 2011–2018, AND PARTNER AT THE STREAMLINER

My first day in, my leg was felt up
I was new to cocktails, a "mixologist" pup
A new kind of bar, I was out of my zone
A misfit bar crew, I wasn't alone
Behind the bar I had spent many years
Pouring shots and slinging draft beers
This place was different, a dark dim-lit room
The vintage my own, not a costume
Notebooks, straw-taste, the flavor in this
Pairs well with cookies, classics with twists
Palate learned this and that to Potato Head drinks

So much rye whiskey, slept zero winks
Downtown was building, some people don't flee
See them shit, piss on the street, and OD
But the neighborhood grew, and safety in pairs
Thieves after hours, (thanks to Matt) I was spared
Seldom again would the groping repeat
Knife in my pocket, taser for the street
Confidence grew, the years I'd been there
Template and practice, not out of thin air
The next detail is to teach what I've learned
The next generation of bartender yearns
To do something different, to make something new
But something's to be said about paying your dues
A timeline, progression, dedication of years
The Varnish. My training. To You, Family, Cheers!

**MAX SEAMAN, VARNISH GENERAL MANAGER,
2011–2018, AND PARTNER AT THE STREAMLINER**

We take our fair share of shit for our rules. Usually it's from assholes who aren't used to being held accountable. Some people simply write us off as pretentious. Sometimes the shit is deserved: sometimes we *are* pretentious, and we definitely have moments when we use the rules as an excuse to be cavalier. But plenty of customers are happy to have an establishment where everyone agrees on how to behave, and there are bartenders, too, who are happy to work at such an establishment. We could make more money and exert less effort working somewhere else, but instead we are here for the gratification of doing a job well and doing it alongside like-minded people. I think the guests can sense the sincerity, and that's what keeps them coming back.

JOSH BITTON, FRIEND OF TWENTY-PLUS YEARS

A few years ago, I went through a really difficult breakup. It was the kind of relationship where marriage and children were very much on the table. When the relationship fell apart, I found myself thrust back into the world of dating. Try not to judge, or judge if you'd like, but there was one week where I had three dates, three nights in a row. Let's say I wasn't the most imaginative date planner—I took all three to the same place. The Varnish. I walked in the first night and was greeted by Anthony.

"Josh, how amazing to see you! Is this the lovely woman I've been hearing so much about?"

A complimentary first round was sent over. The staff fawned all over us, giving space when they sensed it was necessary and engaging in fun, playful, quick bits of conversation, bringing delightful cocktails, making me look like some kind of star.

The second night I walked in with the second date. Mikki greeted us at the door. She had been behind the bar the night before and was fully aware I'd been on a date, but she greeted us as if I hadn't been there in months.

"Oh my God, Josh! It's so good to see you! Who's your date? Wonderful to meet you!"

Night three, I walked in with date number three. Anthony was back on the job at the door, and holy shit—he embraced me as if I was his brother who had been away at war. He kissed my date's hand! He said, "Ahhhh, this must be the lady I've heard so much about!"

Now, maybe they were having a little fun with me. Maybe they knew we all knew the game, but they made my dates feel special. They never intimated anything other than joy at hosting us, making one thing clear: The Varnish wasn't just a bar. It was a home.

CEDD MOSES, CO-OWNER OF THE VARNISH
AND PROPRIETOR OF 213 HOSPITALITY

In late 2007, I bought Cole's, Originators of the French Dip. The place was a mess—rodents wandered the dining room, surveying the seedy customers. The sewer system was clogged from century-old pipes layered with grease. Every time a toilet flushed, the urine would back up into the kitchen. The ceiling was covered in crumbling asbestos that formed a cloud of toxic particles whenever the wall fan was turned on, and they needed the fan on due to the lack of air-conditioning in the kitchen. It was sweltering. Shocking back there.

I wished I hadn't eaten all those dips before we took over.

The first thing we did when getting the keys was call in an abatement crew with hazmat suits to scrape the kitchen clean and remove the asbestos. We junked all the previous kitchen equipment, burned some sage, and boarded it up. I just wanted to pretend that space never existed.

Six months later, I had dinner with Sasha Petraske and Eric Alperin in K-town at Soot Bull Jeep, the smokiest Korean BBQ in town, and suggested working together. They had the best pedigree in the business; Sasha was already a legend. Why would I want to compete?

When they asked if I had anything already licensed, I remembered the toxic cesspool of a space boarded up behind Cole's. I warned them about its history, but they insisted on seeing it.

The Varnish opened fourteen months later. What a miracle . . .

ACKNOWLEDGMENTS

From inception to publication, this book took five long, dramatic years to realize. We could not have done it without the encouragement, kindnesses, ideas, and house keys from the following people:

Publishing: Our agent Rica Allannic, who believed in us when we were lost and has since become not only an agent but an indispensable drinking buddy and friend. Our publisher and editor Karen Rinaldi, who really got us, then pushed us to show the fuck up. Josh Marwell for the amazing title. Rebecca Raskin for generously answering our one million questions; Yelena Nesbit, Brian Perrin, Leah Carlson-Stanisic for the incredible design, and everyone at Harper Wave/HarperCollins who touched our book—we know there are lots of you out there and we hope to meet each and every one of you and thank you in person. To Jim Meehan for introducing us to Rica, Vanessa Vega at Baltz for the PR support, Amy Scattergood and Laurie Ochoa at the *Los Angeles Times*, as well as Laura Compton, Brandi Neal, Sherry Whittemore, Danica Farley, and Mary Woods for all the editorial love lo these many years. Eric Wolfinger, thank you for the best cover photo and shoot ever. Thanks to Alli Phillips for the ice illustrations in chapter 5 and Viviane S. Depigny for the embossed spilled coupe on the hardcover. Mike Clinebell, Andrew Kromelow, and all the contributors in the afterword—we love you.

Friends that held space for us and encouraged us unconditionally: Josh and Mike Bitton, Eric Thorne, Omar Pierre, Rob Mersola, Nicole DaSilva, The Magnificent Seven text thread, Claire van der Boom, Ben Faulks, Ygael Tresser, P. J. Pesce, Joey Box, Dr. Tarek Adra,

Beau Bauman, Fred Berger, Avi Benbasat, Bjorn Lindberg, Courtney Munch, John Munch, Jean Ingold, Modo Yoga LA East, Silas and Amy Halloransteiner, Moon, Cheeles, Katz, Bella Pilar, Coan Nichols, Tallulah Juniper Nichols, Varueka Valentine Nichols, and Nellie Phillips.

Industry friends that encouraged us to keep fighting: Brian Delaney for that very first shift, Simon Ford, Jeffery Morgenthaler, Toby Cecchini and table 99, Alex Day for drives and skydives, The Mixfits and John Lermayer RIP, Gaz Regan RIP, Jeff Hollinger, Johnny Santiago for the late-night motorcycle rides to Boyle Heights Mexican beer bars. BAR Smarts Team: Paul Pacult, Doug Frost, Steve Olson, David Wondrich, Andy Seymour, and Dale DeGroff.

People and places that let us hunker down: Rebekah Bellingham and The Suttle Lodge for hosting us as visiting artists and not murdering us in the avalanche, Freehand Hotel, ACE Hotel, Craftsman Mini-Me, Sally Kim, Dan Thornton and Erin Feil, Carrie Shapiro and Peter Frey, and every bar and café in every country we've been in, our pages and pens soaked in rings from our cocktail glasses—you looked at us funny, but kept us alive and kicking. Thank you.

Bar family: The Chief—Sasha Petraske, past and present staffs at M&H, Little Branch, Middle Branch, Dutch Kills, Fresh Kills, Seaborne, The Everleigh, Silver Lining, Attaboy, Midnight Rambler, The Violet Hour, Hundredweight Ice and Navy Strength Ice Co., Joe Schwartz, Sam Ross, Micky McIlroy, Michael Madrusan, Zara Young, Lucinda Sterling, Toby Maloney, Chad Solomon, Christy Pope, Georgette Moger-Petraske, Nancy Silverton, David Rosoff, the OG Osteria Mozza family, the Delfina and Locanda families, and the OG Spring Lounge family, including Mario Prouini—these stories wouldn't stand without you.

Literary family: Angela Small, Sara Goudarzi, Anthony Rhoades, Vu Tran, Sarah Shanfield. And for all the indelible learnings, special thanks to Mat Johnson, Rick Moody, Helen Schulman, Ted Thompson, Lance Cleland, Tin House, Bread Loaf, Iowa Writers' Workshop, and every workshop member who ever took the time to read and edit and comment—on this book or another—you are all represented here.

Pouring With Heart family: Andrew Abrahamson, Peter Stanislaus for his shutter eye, Brett Winfield, Brian Lenzo, Steven Robbins, Jeannette Rosas, Tim Heller, and everyone at HQ! Friends and business partners—Cedd Moses, Eric Needleman, Mark Verge, Chris Bostick, Richard Boccato, Max Seaman, Mikki Kristola, Adam Weisblatt, Holly Fox, Matt and Mike French, Jean Michel Alperin, Jeff Baker, and Gordon Bellaver. The Lopez-Flores brothers—Carlos, Hector, and David. GMs—Dustin Newsome, Bryan Tetorakis, Dave Fernie, Daniel Zacharczuk, and the teams at The Varnish, Half Step, Bar Clacson, The Slipper Clutch, The Streamliner, The Red Dog Saloon, and Penny Pound Ice. And to all of The Varnish guests and regulars over the years, thank you for putting your trust in us—it would be an empty room without you.

OG family: Powell, Lunder, Weissberg, Linden, Scott, Hawkins, and the Rotkopf families, Michel et Marcelle (Mamie) Depigny, the Alperin clan—Grandma Sally, Joe and Viviane, Jean Michel, Sarah, Simone and Marcel, and the Stoll clan—Marjori, Craig, Annie, Lucy, Andrew, JeanNie, Alina, Graham, and Peter RIP.

And with extra special thanks to lawyer/dad Joe for helping us with the legalese and the indispensable business education over the years, Maman Viviane for all your amazing graphic design ideas, and brother Jean Michel for being a rad dad and our anytime, day-or-night sounding board, love you man. Boundless love to Craig Stoll, our hospitality lifeline, who not only housed and fed us over the years, but answered endless questions about service, engaged in long discussions about vice, and line-edited multiple versions of "Methods and Recipes." And lastly, to Andrew Stoll, our "please finish this line" lifeline, whose bon mots grace the pages of this book, both inside and out, too many times to count. *Unvarnished* would not be as smart or as funny without you.

ABOUT THE AUTHORS

ERIC ALPERIN's first experience behind the stick was at The Screening Room in New York City, followed by the Michelin-starred Lupa and Sasha Petraske's Little Branch. In Los Angeles, he created Osteria Mozza's liquor and cocktail program before opening The Varnish in 2009. The Varnish was a James Beard semifinalist for Outstanding Bar Program from 2013 to 2017 and won Best American Cocktail Bar at Tales of the Cocktail in 2012. Eric was a finalist for Best American Bartender at the Spirited Awards. He is also a co-owner of Half Step in Austin, Texas; the acting director of cocktail bars for Pouring with Heart; and a co-owner of Penny Pound Ice, The Slipper Clutch, Bar Clacson, and The Streamliner at Union Station, all in Los Angeles. He will launch The Red Dog Saloon in Pioneertown, California, in the summer of 2020.

DEBORAH STOLL is a journalist, lyricist, screenwriter, short-story writer, illustrator, and animator. Her essays, articles, and art have appeared in the *Los Angeles Times*, the *Economist*, *LA Weekly*, the *San Francisco Chronicle*, the *Portland Mercury*, and *PUNCH*. Her short stories have been published in *Slake*, *Swivel*, and *Fresh Yarn*. Her songs have been heard on TV shows such as *American Idol*, *Glee*, *CSI: Miami*, *The Vampire Diaries*, and *Pretty Little Liars*; in movies such as *For a Good Time, Call*; and in ads for Ralph Lauren. Her film and TV scripts have been recognized at Sundance, Slamdance, and the Hollywood Screenplay Contest. Long before, during, and long after, she worked in dive bars, cocktail bars, tequila bars, after-hours bars, strip clubs, and nightclubs.